应对气候变化林业行动计划

The Forestry Action Plan to Address Climate Change

国家林业局

二〇〇九年十一月六日

State Forestry Administration，P. R. China

November 6，2009

图书在版编目(CIP)数据

应对气候变化林业行动计划 : 汉英对照 / 国家林业局发布. — 北京 : 中国林业出版社, 2010. 11

ISBN 978 - 7 - 5038 - 5553 - 5

Ⅰ. ①应… Ⅱ. ①国… Ⅲ. ①气候变化 - 林业 - 对策 - 研究 - 中国 - 汉、英 Ⅳ. ①P467②F326. 20

中国版本图书馆 CIP 数据核字(2010)第 210760 号

责任编辑: 洪 蓉

联系电话: 010 - 83228353

出版 中国林业出版社(100009 北京西城区刘海胡同 7 号)

网址 www. cfph. com. cn

E-mail: cfphz@ public. bta. net. cn

发行 新华书店北京发行所

印刷 北京画中画印刷有限公司

版次 2010 年 11 月第 1 版

印次 2010 年 11 月第 1 次

开本 787mm × 960mm 1/16

印张 6. 5

字数 100 千字

定价 20. 00 元

目　录

Contents

第一部分

导 言

自 20 世纪 70 年代以来，以变暖为主要特征的全球气候变化①问题受到了国际社会的日益关注，成为当今国际政治、经济、环境和外交领域的热点问题。

2007 年，政府间气候变化专门委员会②(以下简称 IPCC)正式发布了全球气候变化第四次评估报告，再次用大量数据证实：全球气候变化是一个不争的事实。未来 100 年，全球气候还将持续变暖，将对自然生态系统和人类生存产生巨大影响。导致全球气候变暖的因素包括自然和人为两大类，但主要是由于工业革命以来，人类大量使用化石能源、毁林开荒等行为，向大气中过量排放二氧化碳等

① 《联合国气候变化框架公约》定义的气候变化是指：除在类似时期内所观测的气候的自然变异之外，由于直接或间接的人类活动改变了地球大气的组成而造成的气候变化。

② 政府间气候变化专门委员会(IPCC)是 1988 年由世界气象组织(WMO)和联合国环境规划署(UNEP)共同成立的政府间组织，其作用是在全面、客观、公开和透明的基础上，对全球气候变化进行评估。

温室气体①，导致大气中二氧化碳等温室气体浓度不断增加、温室效应不断加剧的结果。根据 IPCC 全球气候变化第四次评估报告：全球大气中二氧化碳浓度已从工业化前的 280ppm② 增加到了 2005 年的 379ppm，导致全球气温在过去 100 年里约增加了 0.74℃，造成海平面上升、山地冰雪融化、降水量分布和频率及强度发生显著变化、极端天气事件不断增加，并对全球自然生态系统和人类社会可持续发展构成了严重威胁。如果不采取有效措施控制温室气体排放，大气中温室气体浓度将会继续上升，这将使全球平均温度到 2100 年上升 1.4～5.8℃，给全球自然生态系统和人类生存与发展带来不可逆转的影响③。

全球气候变暖也正在对中国产生明显影响。气象观测数据表明：近百年来，中国地表平均气温升高了 0.5～0.8℃。尤其是近 50 年来，中国地表平均气温约增加了 1.1℃，每 10 年约增加 0.22℃，明显高于全球或北半球同期平均地表气温的增温幅度。20 世纪 50 年代以来，中国沿海地区的海平面每年上升 1.4～3.2 毫米，渤海和黄海北部冰情等级下降，西北冰川面积减少了 21%，西藏冻土减薄达 4～5 米，一些高原内陆湖泊水面升高，青海和甘南牧区草地产草量下降。20 世纪 80 年代以来，中国春季物候期提前了 2～4 天，北方干旱受灾面积扩大，南方洪涝加重，海南和广西海域近年来还出现了珊瑚白化现象等④。

为了维护全球生态安全和人类经济社会可持续发展，必须从减

---

① 可导致大气增温的气体统称温室气体，温室气体种类很多。在《联合国气候变化框架公约》下，目前主要将二氧化碳（CO_2）、甲烷（CH_4）、氧化亚氮（N_2O）、氢氟碳化物（HFCs）、全氟化碳（PFCs）、六氟化硫（SF_6）列为管制的温室气体。其中，以二氧化碳为主。

② ppm 指百万分之一。

③ 参考 IPCC 全球气候变化第四次评估报告有关内容。

④ 参考《气候变化国家评估报告》，科学出版社。

缓和适应两个方面积极应对全球气候变暖。减缓主要是指在工业、能源等生产过程中，采取提高能效、降低能耗等措施减少温室气体排放，或者通过发展和保护森林等措施增加对温室气体的吸收，以降低大气中温室气体浓度，减缓全球气候变暖趋势。适应主要是指主动采取措施，增强自然生态系统和人类对气候变暖的适应能力，防止或减少气候变暖的不利影响。

作为发展中国家，中国不承担温室气体量化减排任务。但中国政府深刻认识到，应对全球气候变化、维护全球生态安全，是全人类的共同责任；控制和减少温室气体排放符合建设资源节约型、环境友好型和低耗能、低排放社会的长远发展目标，也是落实科学发展观、实现经济社会可持续发展的内在要求。

多年来，中国政府十分重视应对气候变化工作。早在1990年2月，国务院就专门成立了气候变化对策协调小组，负责协调、制定与气候变化相关的政策和措施，协调小组办公室日常工作由中国气象局承担。1998年后，国务院对气候变化对策协调小组进行了调整，协调小组办公室日常工作改由国家发展和改革委员会承担，负责制定国家应对气候变化的重大战略、方针和政策，协调解决应对气候变化工作中的重大问题。2007年再次调整，成立了国家应对气候变化及节能减排工作领导小组，温家宝总理任组长，国家发展和改革委员会仍承担领导小组办公室日常工作。目前，应对气候变化工作涉及国内20个部门和单位。

为切实履行《联合国气候变化框架公约》(以下简称《公约》)义务，向国际社会阐明中国应对气候变化的政策主张，2007年6月，国务院发布了《中国应对气候变化国家方案》(以下简称《国家方案》)。《国家方案》提出了林业增加温室气体吸收汇、维护和扩大森林生态系统整体功能、构建良好生态环境的政策措施。在《国家方案》公布后不久，就召开了国家应对气候变化及节能减排工作领导小组会议，对贯彻落实《国家方案》进行了总体部署，对各地各部门贯彻落实《国家方案》提出了具体要求。2007年，胡锦涛主席在第

15次亚太经济合作组织(以下简称APEC)会议上，提出了建立“亚太森林恢复与可持续管理网络”的重要倡议，被国际社会誉为应对气候变化的“森林方案”。国家林业局积极行动，于2007年7月成立了国家林业局应对气候变化和节能减排工作领导小组及其办公室，积极开展工作，并着手组织编制《应对气候变化林业行动计划》(以下简称《林业行动计划》)，以贯彻落实《国家方案》中赋予林业的任务，指导各级林业部门开展应对气候变化相关工作。

第二部分

林业与气候变化

一、森林在应对全球气候变化中具有独特作用

森林是陆地最大的储碳库和最经济的吸碳器。据 IPCC 估算：全球陆地生态系统中贮存了约 2.48 万亿吨碳，其中 1.15 万亿吨碳贮存在森林生态系统中。森林通过光合作用吸收二氧化碳，放出氧气，把大气中的二氧化碳固定在植被和土壤中，这个过程被称为碳汇。科学研究表明：林木每生长 1 立方米，平均约吸收 1.83 吨二氧化碳，放出 1.62 吨氧气。全球森林对碳的吸收和储量占全球每年大气和地表碳流动量的 90%。森林的碳汇功能和其他许多重要的生态功能一样，对维护全球生态安全、气候安全发挥着重要作用。

森林锐减是导致全球气候变化的重要因素之一。全球气候变暖主要是大气中二氧化碳等温室气体浓度升高导致温室效应的结果。大气二氧化碳浓度升高有两个主要原因：一是大规模燃烧化石能源，排放二氧化碳；二是全球森林锐减，释放二氧化碳。目前，全球森林已从人类文明初期的约 76 亿公顷减少到 38 亿公顷。联合国《2000 年全球生态展望》指出，全球森林减少了 50%。现在，森林减少的趋势仍在继续。联合国粮食与农业组织（以下简称 FAO）的报

告显示：2000～2005 年，全球年均毁林[①]面积为 730 万公顷。

恢复和保护森林是缓解全球气候变化最根本的措施之一。IPCC 在 2007 年发布的全球气候变化第四次评估报告中指出：与林业相关的措施，可在很大程度上以较低成本减少温室气体排放并增加碳汇，从而缓解气候变化。围绕《京都议定书》第二承诺期谈判，许多国家和国际组织都在积极倡导通过恢复和建设森林生态系统，来缓解气候变化。

同时，气候变化又严重影响了森林。森林生长自始至终受到光照、温度、水分和风等自然因素的影响，这些因素都和气候有着紧密联系。因此，需要积极提高森林适应气候变化的能力，减少气候变化对森林的不利影响，维持森林良好的生态功能。

二、中国林业建设成就及对减缓全球气候变化的贡献

中国政府历来高度重视发展和保护森林。自 1978 年以来，先后在三北（东北、西北、华北）、沿海、平原、长江中上游、太行山、京津周围、淮河和太湖流域、珠江流域、辽河流域等地区实施了一系列区域性防护林体系建设工程。1998 年调整林业发展布局后，启动试点并相继实施了天然林保护、退耕还林、京津风沙源治理、三北和长江等地区防护林建设、速生丰产林基地建设以及野生动植物保护六大工程。截至 2008 年，六大工程完成造林面积 5153.74 万公顷（含封山育林 1475.38 万公顷）。总投资 2781.26 亿元，其中，国家投资 2416.36 亿元[②]。1981 年以来，中国持续开展了全民义务植树运动。截至 2008 年底，全国共有 115.2 亿人次义务植树 538.5 亿株，城市绿化覆盖率由 1981 年的 10.1% 提高到 35.29%，人均公共

---

① 《公约》下所涉及的毁林是指有林地转化为非林业用地的情况。如林地转化为农地、牧地或城市基础设施建设用地等。

② 引自《中国林业统计年鉴》（2008），（国家林业局），中国林业出版社。

绿地面积由3.45平方米提高到8.98平方米，促进了城乡绿化，改善了人居环境①。

为了保护森林，中国先后出台了9部林业法律、15部林业行政法规、43部林业部门规章、300余件地方性法规规章，形成了以《森林法》、《野生动物保护法》、《防沙治沙法》为核心的森林资源保护法律体系和以林政管理为主体，资源监测、监督为两翼的森林资源管理体系。多次实施了打击乱砍滥伐、乱征乱占林地、湿地等违法犯罪行为的专项行动。2001~2008年，全国共查处各种破坏森林资源案件331.7万起。同时，还加大了对森林火灾和病虫害的防控和自然保护区建设力度。目前，全国已建有各种类型自然保护区2531个，占国土面积的15.2%②。

通过采取一系列发展和保护森林资源的措施，中国森林面积和蓄积量实现了持续增长。据第六次全国森林资源清查(1999~2003年)：中国森林面积已达1.75亿公顷，森林覆盖率为18.21%，占世界森林面积的4.5%，列世界第五；森林蓄积量124.56亿立方米，占世界总量的3.2%，列世界第六。人工林保存面积0.54亿公顷，约占全球人工林总面积的1/3，居世界首位③。

中国林业建设成就得到了国际社会的广泛认可。据FAO《2005世界森林资源状况》评估报告：2000~2005年，在全球森林资源继续呈减少趋势的情况下，亚太地区森林面积出现了净增长。其中，中国森林资源的增长在很大程度上抵消了其他地区的高采伐率④。FAO《2009世界森林资源状况》评估报告再次肯定了中国森林资源持

① 引自《2008年中国国土绿化状况公报》，全国绿化委员会办公室2009年3月11日发布。

② 参考《中国林业发展报告》(2001~2008)有关部分，国家林业局主编，中国林业出版社。

③ 参考《中国森林资源》，雷加富主编，中国林业出版社。

④ 《2005世界森林资源状况》，FAO。

续增长的成就①。

中国森林面积和蓄积量的持续增长，在增加中国木材自给和改善中国生态环境的同时，也吸收固定了大量的二氧化碳。据专家估算：1980～2005年，中国通过持续不断地开展植树造林和森林管理活动，累计净吸收二氧化碳46.8亿吨，通过控制毁林，减少二氧化碳排放4.3亿吨，两项合计51.1亿吨②，对减缓全球气候变暖作出了重要贡献。

森林碳储量反映了森林生态系统吸收固定二氧化碳的总体情况。不同估算方法会导致估算结果有较大差异。方精云院士等利用1977～2003年全国森林资源清查数据进行分析表明：自20世纪70年代末以来，中国森林植被碳库呈现显著增加趋势，单位面积的森林碳密度已由20世纪80年代初期每公顷36.9吨碳增加到2003年的41吨碳。

对中国森林碳汇未来变化趋势的研究结果虽然因研究方法不同而有差异，但总体趋势是：1990～2050年，中国森林的碳储量将会逐步增加。

三、中国林业减缓气候变化途径和潜力初步分析

林业在减缓气候变化中的作用主要是通过增汇、减排、储存、替代四个途径来实现。具体措施包括通过植树造林、植被恢复、可持续经营森林措施增加森林碳吸收；通过合理控制采伐、减少毁林、防控森林火灾与病虫害，减少源自森林的碳排放；通过增加木质林产品使用，延长木材使用寿命，扩大木质林产品碳储量；利用木质林产品和林木果实，转化为能源以部分替代化石能源，如森林采伐和加工剩余物能源化利用、林木果实转化生物柴油等，将有助于减少化石能源使用量，从而减少碳排放。中国发展林业生物质能

① 《2009世界森林资源状况》，FAO。

② 引自《中国应对气候变化国家方案》。

源具有较大潜力，应积极开发和利用。

（一）通过植树造林，扩大森林面积，增加碳汇。与主要发达国家和一些发展中国家相比，中国森林覆盖率较低。中国尚有0.57亿公顷宜林荒山荒地、0.54亿公顷左右的宜林沙荒地、相当数量的25度以上的陡坡耕地和未利用地都可用于植树造林。同时，通过提高现有林地使用率，发展农田林网等途径，扩大中国森林面积尚有较大空间。根据《中共中央　国务院关于加快林业发展的决定》中所确定的林业中长期发展目标，到2050年，中国森林覆盖率将由现在的18.21%提高到26%以上。届时，森林碳储量将会得到较大提高。

（二）通过提高现有森林质量增加碳汇。中国现有森林资源平均蓄积量约为每公顷84立方米，每公顷林分年均生长量约为3.55立方米，大多数森林属于生物量密度较低的人工林和次生林。专家分析：中国现有森林植被资源的碳储量只相当于其潜在碳储量的44.3%。因此，通过合理调整林分结构，强化森林经营管理，在现有基础上，完全有可能将单位面积林分生长量提高1倍以上，从而大大增加现有森林植被的碳汇能力。

（三）通过加强森林保护，减少森林碳排放。首先，通过严格控制乱征乱占林地等毁林活动，减少源自森林的碳排放。中国历次森林资源清查结果表明：中国每年因乱征乱占林地而丧失的有林地面积约100万公顷左右。因此严格控制乱征乱占林地等毁林行为，对控制碳排放具有较大潜力。同时，在森林采伐作业过程中，通过采取科学规划、低强度的作业措施，保护林地植被和土壤，可减少因采伐对地被物和森林土壤的破坏而导致的碳排放。其次，发生森林火灾和病虫害都会导致储存在森林生态系统中的碳在短时间内释放到大气中。因此，通过强化对森林中可燃物的有效管理，建立森林火灾、病虫害预警系统等措施，有效控制森林火灾和病虫害发生频率和影响范围，减少森林碳排放。

（四）通过保护湿地和控制林地水土流失，减少温室气体排放。首先，湿地土壤中储存着大量的有机碳，若遭受破坏，其储存的有机碳就会分解，并向大气中排放二氧化碳等温室气体。中国现有100公顷以上的各类湿地总面积3848万公顷。由于经济社会发展，大量湿地退化或被占用。加大湿地保护力度，可以减少因湿地破坏而导致的温室气体排放。其次，森林土壤中也储存了大量有机碳，约占整个森林生态系统碳储量60%以上。通过加大生物措施，控制林地水土流失，有助于保护林地土壤，促进和加速森林土壤发育，促使非森林土壤转化为森林土壤，提高森林土壤固碳能力。

（五）通过发展林木生物质能源替代化石能源，减少碳排放。林木生物质原料通过直接燃烧、木纤维水解转化为乙醇、热解气化以及利用油料能源树种的果实生产生物柴油等途径，都可以部分替代化石能源，减少温室气体排放。据统计，中国每年有可以能源化利用的森林采伐和木材加工废弃物3亿多吨，如果全部利用，约可替代2亿吨标准煤。同时，利用现有宜林荒山荒地和盐碱地、矿山复垦地等难利用地，还可定向培育一部分能源林，扩大林木生物质替代化石能源的比例，有利于减少中国温室气体排放总量。

（六）通过增加木材使用、延长使用寿命，增加木质林产品碳储量。木材在生产和加工过程中所耗能源，大大低于制造铁、铝等材料导致的温室气体排放。用木材部分替代能源密集型材料，不但可以增加碳贮存，还可以减少使用化石能源生产原材料所产生的碳排放。研究表明：用1立方米木材替代等量水泥、砖等材料，约可减排0.8吨二氧化碳当量，还节约了能源，又减少污染。木制品只要不腐烂、不燃烧，都是重要碳库。专家初步测算：1961～2004年间，中国木制品碳储量约达12亿～18亿吨二氧化碳当量，这是林业对减缓气候变化的重要贡献。

四、气候变化对中国林业发展的影响

发展林业有助于减缓气候变化；而气候变化会引起温度、湿度、生长季节、降水和蒸发等气候因子的变化，特别是极端天气发生频率的增加，会对林业发展构成现实和潜在影响。根据中国《气候变化国家评估报告》，气候变化对中国森林和林业发展的主要影响是：未来气候的持续变暖，将会对中国森林生态系统稳定性、结构和功能产生不利影响。

从植被分布看，将可能导致中国东部亚热带、温带地区的植被普遍北移，物候期提前，主要造林树种北移，并对生物多样性构成威胁。从森林生产力看，气候变暖虽然可能会使中国森林生产力呈现不同程度的增加，但不会改变中国森林生产力目前的地理分布格局。热带、亚热带大部分地区的森林生产力增幅只有1%；寒温带和西南亚高山林区森林生产力增幅可达10%；而暖温带、温带森林生产力增幅可能为2%～8%。从动植物生境看，一些珍稀树种如秃杉、珙桐的分布区和大熊猫、滇金丝猴、藏羚羊等濒危野生动物栖息地将缩小，一些适应能力差的物种将加速灭绝。从森林灾情看，气候变化会导致中国区域气候特征和规律发生异常变化，加剧森林火灾发生频度和强度，如雷击火发生次数增加，防火期延长，极端火险条件和严重程度加剧等，将直接危害森林生长，并可能破坏森林生态系统结构和功能。同时，气候变暖还会使森林病虫害分布区向北扩大，发生期提前，世代数增加，发生周期缩短，发生范围和危害程度加大。同时，还会加重外来入侵病虫害危害程度，并通过影响病原体存活和变异以及媒介昆虫孳生分布和流行病学特征等，导致带菌者和疾病分布的纬度上移，对野生动物生存繁衍造成不利影响。从旱涝变迁看，在未来气候变暖情形下，中国西部沙漠和草原将可能会略有退缩而被草原和灌丛取代，但气候变暖将加剧冻土退化、冰川退缩和水资源短缺，进一步影响内陆河流。极端干旱和亚湿润干旱区将大幅度增加，全国荒漠化和水土流失总面积将呈扩大趋势。从湿地功能看，气候变暖将导致北方河流断流、湖泊萎缩、水库蓄水量减少、海平面上升，进一步导致湿地面积缩减，功能下降，沿海地区红树林生态系统将受到较大损害。

由于中国森林资源总量不足，随着工业化、城镇化进程加快，在气候变暖情景下，林地、湿地、沙地保护压力加大，将给植树造林和生态恢复带来严重挑战。因此，必须采取有效措施增强森林适应气候变化的能力。森林生态系统适应气候变化能力的提高，也有助于进一步增强森林减缓气候变化的能力。

第三部分

应对气候变化的国际进程与林业

一、应对气候变化国际进程中的林业问题

1992 年，在巴西召开了首届联合国环境与发展大会，通过了《公约》，确立了“将大气中温室气体的浓度稳定在防止气候系统受到危险的人为干扰的水平上”的目标。《公约》于 1994 年正式生效。中国是《公约》缔约方之一。

为了实现《公约》目标，各国都要履行《公约》义务，积极采取减缓和适应措施，不断增强应对气候变化的能力。由于排放到大气中的温室气体主要源自发达国家的历史排放，本着“共同但有区别的责任”的原则，发达国家缔约方应率先减排。1997 年 12 月，在日本京都召开的《公约》第三次缔约方大会上通过了《京都议定书》，首次以法律形式规定《公约》附件一国家（包括主要工业化国家和经济转轨国家，统称发达国家）在 2008 ~2012 年，要把本国温室气体排放量在 1990 年的基础上平均减少 5.2%。

为了帮助附件一国家完成他们在《京都议定书》中承诺的减排任务，《京都议定书》规定发达国家可借助三种灵活机制来履约，这三种灵活机制是排放贸易、联合履约和清洁发展机制。其中，排放贸易和联合履约实质上是发达国家之间的温室气体排放权交易，而清

洁发展机制则是指发达国家可以通过和发展中国家合作开展减排或增汇项目，以获得核证减排量，用于抵消《京都议定书》为发达国家规定的减排量。清洁发展机制的实质是发达国家向发展中国家购买温室气体排放权，要求发达国家要以输入资金和技术转让等形式，在发展中国家施项目，在从发展中国家获得温室气体排放权的同时，要推进发展中国家经济社会可持续发展。

通过林业活动增加碳汇、减少排放，被作为履行《公约》和《京都议定书》的重要措施，在《京都议定书》通过后，各缔约方就如何通过林业活动来帮助发达国家完成减排任务进行了长时间谈判，最终形成了一系列缔约方大会决定。归纳起来，有两种方式：一是发达国家可以利用本国1990年以来的相关林业活动产生的碳汇来抵消其2008~2012年的温室气体排放量；二是发达国家可以利用清洁发展机制，购买在发展中国家实施的造林、再造林项目产生的碳汇，以部分抵消其在2008~2012年的温室气体排放量。按照缔约方大会有关决定，发达国家利用林业碳汇可完成《京都议定书》为本国规定的减排任务的20%~30%。由于林业碳汇成本较低，大大减轻了发达国家履行《京都议定书》减排承诺的压力。

与此同时，热带地区的一些发展中国家长期以来面临着严重的毁林困扰。IPCC评估报告表明：全球毁林排放的二氧化碳多于交通部门，是位居能源、工业之后的全球第三大温室气体排放源，约占全球温室气体总排放量的20%左右。经过一系列谈判，2007年底在印度尼西亚巴厘岛召开的《公约》第13次缔约方大会，将减少发展中国家毁林和森林退化导致的碳排放等相关内容纳入了《巴厘行动计划》。如何发挥发展中国家林业在减缓气候变化中的重要作用已成为未来全球减缓气候变化共同行动的重要组成部分。

由于林业在应对气候变化中的特殊作用，在应对气候变化的国际进程中，不论是帮助发达国家完成其承诺的量化减排指标，还是进一步推进发展中国家参与减缓全球气候变化行动，林业始终都承担着重要任务。可以预见，这将给林业发展带来诸多挑战和机遇。

二、气候变化给林业发展带来的挑战

（一）气候变化将对中国森林生产力、物种分布和生态系统稳定性产生重要影响。如果不能很好地防控气候变化对森林的不利影响，森林不仅不能起到减缓气候变化的作用，还会加剧气候变暖趋势，进而影响森林自身的健康发展。近年来，气候变暖导致中国许多地区的森林火灾和病虫害发生频率和强度呈加剧趋势，西部干旱和半干旱地区水资源短缺状况日趋严重等。总体上，气候变化将加大中国森林资源保护和发展的难度。

（二）气候变化将加剧土地类型和不同利用方式间的矛盾。研究表明：气候变化可能对中国农业生产布局和结构产生很大影响，导致种植业生产能力下降。在人口数量增加的情况下，将意味着有更多的森林或林业用地面临被毁或征占用于粮食和畜牧业，势必加剧不同土地利用方式间的矛盾，这将加大林业部门管理森林和林地的难度，对通过扩大森林面积增加碳汇构成了制约。

（三）气候变化对全球木质和非木质林产品以及森林生态服务的供给产生影响。大量研究表明：虽然通过林业措施减缓气候变化可带来多重效益，有助于降低减缓气候变化的成本，但也会导致土地利用格局的变化。在应对气候变化背景下，如何平衡森林提供林产品和包括增加碳汇在内的各种生态产品的需求，并为当地林业经营者提供持续有效的激励，就需要对中国现行林业政策、体制和机制进行改革和创新。

（四）随着《公约》谈判进程的不断深入，减少发展中国家毁林和森林退化造成的碳排放等行动将逐步纳入减缓气候变化的范畴，势必增加森林采伐和利用的成本，将在一定程度上加大中国进口木材成本，对中国利用境外森林资源形成制约。这对中国调整完善林业相关政策措施，提高木材自给能力，提出了新要求。

三、应对气候变化给林业带来的发展机遇

（一）IPCC 第四次评估报告认为：林业是当前和未来 30 年乃至更长时期内，技术和经济可行、成本较低的减缓气候变化重要措施，可以和适应形成协同效应，在发挥减缓气候变暖作用的同时，带来增加就业和收入、保护水资源和生物多样性、促进减贫等多种效益。在气候变化大背景下，宣传林业减缓气候变化的作用，有助于促进全社会重新认识森林价值和林业工作的重要性，形成全社会重视林业、发展林业的良好氛围。

（二）《公约》和《京都议定书》下的创新机制，为促进林业发展提供了新机遇。尤其是基于排放权交易的碳市场的产生和发展，有助于对碳排放行为进行市场定价，通过价格机制既能约束排放主体的排放行为，又能降低全球温室气体减排总成本。林业碳汇是全球碳交易的组成部分。通过碳市场，开展碳汇交易，实现林业碳汇功能和效益外部性的内部化。从近期看，有助于将森林生态效益使用者和提供者的利益有机地结合起来，进一步完善生态效益补偿机制。从长远看，则有助于推进林业发展投融资机制的改革和创新。

（三）根据《巴厘行动计划》，减少发展中国家毁林和森林退化导致的碳排放，以及通过森林保护、可持续经营和造林增加碳汇已成为 2012 年后发展中国家在更大程度上参与减缓气候变化行动的重要内容。发展中国家在这方面能否采取有效行动，将取决于发达国家在多大程度上为发展中国家提供资金和技术支持等。因此，将林业纳入应对气候变化国际和国内进程，将为林业发展提供新的机会。

（四）充分发挥林业在应对气候变化中的作用，不仅涉及造林、森林经营，还涉及通过发展林木生物质能源替代化石能源和利用生物质材料替代化石能源生产的原材料等方面。如利用油料能源林生产的果实榨油可转化为生物柴油；利用定向培育的能源林、林区采伐剩余物、木材加工废料等可直燃发电或供热；利用林木半纤维素转化为乙醇燃料可作为第二代生物燃料；利用木材可直接替代部分

化石能源生产砖、钢材、铝材、玻璃等原材料。这些不仅可以大大降低温室气体排放，也可以为促进林业乃至经济社会可持续发展提供新的增长点。

（五）按照《京都议定书》规定，中国正在积极参与实施清洁发展机制下的造林、再造林项目。这不仅为中国引入了一定数量的造林资金，也为熟悉相关国际规则，开展碳汇计量、监测、核查、交易等提供了经验，有助于增强参与实施碳汇项目的能力，为借助市场机制进一步完善森林生态价值补偿制度，扩大造林绿化资金渠道，加快中国造林绿化步伐提供借鉴。

总之，在气候变化大背景下，林业发展既面临着重大挑战，也面临着战略机遇。气候变化将进一步促进各国政府更多地关注林业，加快林业管理制度改革和林业发展机制创新。主动抓住机遇，积极应对挑战，将给各国林业发展带来新动力。

第四部分

林业应对气候变化的指导思想、基本原则和主要目标

一、指导思想

以科学发展观为指导，按照《国家方案》提出的林业应对气候变化的政策措施，结合林业中长期发展规划，依托林业重点工程，扩大森林面积，提高森林质量，强化森林生态系统、湿地生态系统、荒漠生态系统保护力度。依靠科技进步，转变增长方式，统筹推进林业生态体系、产业体系和生态文化体系建设，不断增强林业碳汇功能，增强中国林业减缓和适应气候变化的能力，为发展现代林业、建设生态文明、推动科学发展作出新贡献。

二、基本原则

（一）坚持林业发展目标和国家应对气候变化战略相结合。确定林业发展目标要充分考虑国家应对气候变化战略，把增强林业经济、生态和社会功能与增强森林减缓和适应气候变化的能力有机统一起来。在制定各级应对气候变化战略和政策中，将林业作为重要措施加以重视和支持。

（二）坚持扩大森林面积和提高森林质量相结合。一方面要继续

通过扩大森林面积，加大退化湿地恢复和沙化土地的治理力度，增加林业碳汇；另一方面，要努力提高单位面积森林的年生长量和固碳能力，通过科学经营森林，将生物量和碳密度较低的林分，逐步转变为生物量和碳密度较高的林分，全面增强中国现有森林的固碳能力和相关的综合效益。

（三）坚持增加碳汇和控制排放相结合。既要通过扩大森林面积，加大退化湿地恢复和沙化土地的治理力度，以及提高现有森林质量，增加林业碳汇，又要积极采取措施，保护森林、湿地和荒漠生态系统的资源，防止森林、湿地和荒漠生态系统遭受破坏而导致储存在这些生态系统中的碳被重新排放到大气中。

（四）坚持政府主导和社会参与相结合。既要发挥政府在推进林业发展中的主导地位，又要继续坚持全民参与、全社会办林业的做法。通过多种形式，调动企业、团体、组织和个人积极参与植树造林和保护森林、增加碳汇等应对气候变化的行动。

（五）坚持减缓与适应相结合。既要通过增加森林碳汇、减少森林碳排放来增强林业减缓气候变化的作用，又要高度重视林业适应气候变化的能力，将适应作为增强林业减缓气候变化的基础加以重视，使林业减缓和适应气候变化之间形成协同效应。

三、主要目标

（一）总体目标。推进宜林荒山荒地造林，扩大湿地恢复和保护范围，加快沙化土地治理步伐。继续实施好天然林保护、退耕还林、京津风沙源治理、速生丰产用材林、防护林体系建设工程和生物质能源林基地建设；努力扩大森林面积，增强中国森林碳汇能力；重视和加强森林可持续经营，提高单位面积林地的生产力，增强单位面积森林的年生长量和固碳能力；采取有力措施，加大森林火灾、森林病虫害、野生动物疫源疫病防控力度，合理控制森林资源消耗，打击乱砍滥伐和非法征占用林地和湿地行为，切实保护好森林、荒漠、湿地生态系统和生物多样性，减少林业排放。积极强化林业生产中的适应性管理措施，努力提高林业适应气候变化能力，充分发挥林业在应对气候变化国家战略中的作用。

（二）阶段目标[①]。分三个阶段性目标：

（1）从现在起到2010年，年均造林（含封山育林）面积400万公顷[②]以上，全国森林覆盖率达到20%，森林蓄积量达到132亿立方米。生态环境特别恶劣的黄河、长江上中游水土流失重点地区以及严重荒漠化地区的治理初见成效。国家重点公益林保护面积达到0.51亿公顷，50%的自然湿地得到有效保护，人工林良种使用率达到50%。届时，森林碳汇能力将得到较大增长。

（2）2011~2020年，年均造林（含封山育林）面积500万公顷以上，全国森林覆盖率增加到23%，森林蓄积量达到140亿立方米。新增沙化土地治理面积占适宜治理面积的50%以上，约1.1亿公顷国家重点公益林得到有效保护，60%以上的自然湿地得到良好保护，人工林良种使用率达到65%。实现2020年森林面积比2005年增加4000万公顷，森林蓄积量比2005年增加13亿立方米的目标。届时，中国森林生态系统整体固碳功能将进一步增强，森林碳汇能力将得到进一步提高。

（3）到2050年，比2020年净增森林面积4700万公顷，森林覆盖率达到并稳定在26%以上，典型生态系统得到良好保护，适宜治理的沙化土地基本得到治理，全国自然湿地得到有效保护、恢复和合理利用，全国人工林基本实现良种化，林业发展重点转向全面开展森林可持续经营阶段，森林碳汇能力保持相对稳定。

---

① 参考《中国可持续发展林业战略研究总论》和《林业发展"十一五"和中长期规划》相关部分。

② 参考1999~2003年第六次全国森林资源清查结果。

第五部分

林业应对气候变化的重点领域和主要行动

为了充分发挥林业在应对气候变化中的独特作用，根据中国林业可持续发展战略、林业中长期发展规划以及《国家方案》对林业发展的总体要求，从提高林业减缓和适应气候变化两个方面确定了以下重点领域和主要行动。

一、林业减缓气候变化的重点领域和主要行动

领域一：植树造林

行动1：大力推进全民义务植树。各级政府要继续按照全国人大《关于开展全民义务植树运动的决议》和国务院《关于开展全民义务植树运动的实施办法》，把开展好全民义务植树纳入重要议事日程，层层落实领导责任制。要认真落实属地管理制度，强化乡镇政府和城市街道办事处组织实施义务植树的职能，确保适龄公民履行义务。要加强对各部门、各单位履行义务情况的检查和监督，探索和丰富义务植树活动的实现形式，努力提高全民义务植树尽责率。要进一步调动各部门、各单位和社会各界参与造林绿化的积极性，重点抓好城市、绿色通道、村庄和校园绿化工作。

行动2：实施重点工程造林，不断扩大森林面积。天然林保护工程要切实巩固现有建设成果，继续限制项目区内天然林的商品性采伐，加强项目区内宜林荒山荒地造林，对现有天然林实施全面有效保护。

退耕还林工程要进一步加强检查验收、政策兑现、确权发证、效益监测和后期管护工作，落实基本农田建设和相关成果巩固配套政策，搞好工程质量评价，在巩固工程建设成果的基础上稳步有序推进。

京津风沙源治理工程要加强项目区内荒山荒地造林和沙化土地治理，大力推广先进实用技术与治理模式，认真执行禁止滥开垦、滥放牧、滥樵采制度，加强林分抚育和管护工作，切实巩固工程治理成果。

三北防护林体系建设工程要突出防沙治沙和水土流失治理，构建完善的三北地区农田防护林体系，重点抓好区域性防护林体系和示范区建设，进一步调动全社会力量，努力建设生态经济型防护林体系，构筑稳固的北方地区生态屏障。

长江、珠江、沿海防护林和太行山、平原绿化工程要根据不同区域的治理要求采取不同措施。长江防护林要加强对鄱阳湖、洞庭湖流域和三峡、丹江口库区水土流失治理，搞好低效林改造，巩固建设成果；珠江防护林要突出石漠化治理，加大封山育林力度，建设高效水源涵养林和水土保持林；沿海防护林要以现有森林资源为基础，进一步拓宽和完善沿海基干林带，重点加强红树林保护、恢复和管理力度，力争实现全面恢复和保护沿海红树林区及其湿地环境，提高沿海地区抵御海洋灾害的能力，最大限度地减少海平面上升造成的社会影响和经济损失；太行山绿化要着眼于建设华北平原的生态屏障，搞好河源区水源涵养林建设和保护；平原绿化要重点建设华北、东北等平原地区的高标准农田防护林，加快村屯绿化、四旁植树、平原农田林网更新改造步伐，抓好绿色通道工程建设。

重点地区速生丰产林基地建设工程要积极鼓励林产加工等用材

企业发展原料林基地，建立大径材培育基地和竹林培育基地，推动林纸、林板一体化建设。构建速生丰产用材林绿色产业带及国家木材储备基地。组织编制和落实省级工程建设规划，完善速丰林技术标准，提高工程建设质量。逐步增强人工用材林的碳汇能力。

行动3：加快珍贵树种用材林培育。在适宜地区，结合工业原料林基地、天然林保护和退耕还林工程，积极建立珍贵树种用材林培育基地。针对天然林中的珍贵树种资源进行高效栽培和可持续利用，有目的地培育珍贵天然用材林和其他用途的森林资源。优化珍贵树种用材林培育技术，选择优良林型，合理调控林分密度，优化林分结构，提高林分光能利用率和林分生产力。

领域二：林业生物质能源

行动4：实施能源林培育和加工利用一体化项目。尽快实施《全国能源林建设规划》。一是要充分利用山区、沙区等边际土地和宜林荒地，大力发展小桐子、黄连木、文冠果、光皮树等木本油料树种，建设一批以生产生物柴油为目的的油料能源林示范基地，重点抓好与中国石油天然气集团公司等合作的生物质能源项目。二是要充分利用退耕还林、防沙治沙工程发展起来的灌木资源，以及主伐、间伐、木材加工剩余物，加工成用于直燃发电或供热的高效固体成型燃料。三是要积极支持开发生物质能高效转化发电技术、定向热解气化技术和液化油提炼技术，逐步形成原料培育、加工生产、市场销售、科技开发的“林能一体化”格局。

领域三：森林可持续经营

行动5：实施森林经营项目。以提高现有森林年生长量为目标，制定和实施“人工商品林经营规划”。以提高森林生态功能为目标，制定和实施“全国重点公益林经营规划”。在国家和省级层面上，重点落实分区施策、分类管理，按照不同自然、地理特点和经济状况进行区划，合理划定公益林和商品林。针对不同区域、不同类型森林采取相应的管理政策。在县级层次上，重点开展森林经营规划，明确各类森林培育方向和经营模式。在经营单位层面上，重点编制和实施森林经营方案，将不同经营措施落实到山头地块，把主要经营任务落实到年度。在林分经营层面上，充分运用现代森林经营技术和手段，最大限度地提高林地生产力，使不同林分的目标效益最大化。在实施森林可持续经营项目中，要建设一批示范点，探索不同条件下的森林经营模式，积极推广森林可持续经营指南，建立符合中国林业发展特点的森林可持续经营指标体系。认真执行《森林经营方案编制与实施管理办法》和《生态公益林抚育技术规程》等技术规定，强化森林健康理念，不断提高森林生态系统的抗逆性和稳定性，充分发挥现有森林资源的碳汇潜力。

行动6：扩大封山育林面积，科学改造人工纯林。封山育林是一种成本较低、活动过程中温室气体排放较低的森林恢复方式。要尽可能地扩大封山育林面积，加快次生林恢复的进程。要加强对现有人工林的经营管理，对人工纯林进行适度的“抽针补阔”，逐步解决“过密、过疏、过纯”问题。尽可能避免长期在相同的立地上多代营造针叶纯林。要根据未来气候变化情景，尽量避免在中国气候带交错区域营造大面积人工纯林，努力增强人工纯林抗御极端和灾害性天气的能力。

领域四：森林资源保护

行动7：加强森林资源采伐管理。严格执行林木采伐限额制度，对公益林和商品林采伐实行分类管理。公益林要完善森林生态效益补偿基金制度，确保稳定高效地发挥其生态效益。商品林尤其是速生丰产用材林和工业原料林，要依法放活和优先满足其采伐指标。要修订《森林采伐更新管理办法》和相关采伐作业规程，促进森林资源利用管理的科学化和法制化。要在科学区划的基础上，针对不同区域，按照林业发展布局和森林主体功能要求，实行不同的采伐管理模式，将森林采伐管理与分区施策以及森林经营方案结合起来，做到有效保护和科学经营森林资源。

行动8：加强林地征占用管理。科学编制“林业发展区划”和“全国林地保护利用规划纲要”，明确不同区域林业发展的战略方向、主导功能和生产力布局。强化林地保护管理，把林地与耕地放在同等重要位置，采取最严格的保护措施，建立和完善林地征占用定额管理、专家评审、预审制度。实施林地保护利用规划和林地用途管制。严格执行征占用林地的植被恢复制度，做到林地占补平衡。最大限度地减少林地征占用造成的碳排放。

行动9：提高林业执法能力。逐步建立起权责明确、行为规范、监督有效、保障有力的林业行政执法体制，充分发挥各级林业主管部门及其森林公安、林政稽查队、木材检查站、林业工作站以及广大护林员队伍的作用，加强森林资源保护。要加大执法力度，依法严厉打击各类破坏森林资源的违法行为。对森林资源管理混乱、破坏严重的地区，要定期或不定期地开展专项整治行动。对一些重大、典型案件要一查到底，决不姑息。

行动10：提高森林火灾防控能力。坚持“预防为主、积极消灭”的原则，采取综合措施，全面提升森林火灾综合防控水平，最大限度地减少森林火灾发生次数，降低火灾损失。认真贯彻落实《森林防火条例》，加强以森林防火指挥为核心的应急管理组织体系建设。组织实施《全国森林防火中长期发展规划》，改善森林防火装备和基础设施建设水平，大力加强森林消防专业队伍建设，加快生物防火隔离带建设步伐，提高火灾应急处置能力。加强火险预报，建立森林火险预警体系和分级响应机制。加强防火宣传、火源管理、隐患排查等防范措施，不断提高全民防火意识，减少人为火灾的发生，推动森林防火由盲目设防、应急扑救为主向主动设防、有准备扑救的转变，实现“打早、打小、打了”。加大对森林防火新技术、新装备的研发引进力度，逐步扩大灭火飞机、全道路运兵车等大型灭火装备的应用范围，增强森林大火的扑救能力。加强与周边国家的联系与协商，建立突发自然火灾紧急互助机制。

行动11：提高森林病虫鼠兔危害的防控能力。坚持“预防为主、科学防控、依法治理、促进健康”的方针，做好森林病虫鼠兔危害的防治工作。修订《森林病虫害防治条例》。加强和完善应急管理和对松材线虫病、美国白蛾、椰心叶甲、红脂大小蠹、松突圆蚧、杨树蛀干害虫等重要外来有害生物和有重要影响的本土病虫害的除治。制定和实施全国林业有害生物防治2008~2015年建设规划。全面加强森林病虫鼠兔危害的监测预报工作以及1000个国家级中心测报点的建设和管理。加强和国家气象主管部门的合作，增强监测预报的科学性、时效性和准确性。加强森林病虫害的检疫执法，与海关部门密切合作，严防外来有害生物的入侵。

领域五：林业产业

行动12：合理开发和利用生物质材料。要抓好生物质新材料、生物制药等开发和利用工作。制订生物质材料开发利用规划。落实《林业产业政策要点》，避免低水平重复，控制高耗能高污染企业，促进林业循环经济发展。要在巩固木材传统应用领域的基础上，通过木质产品性能改良，积极扩大木材在建筑、包装、运输和能源等领域的应用，大力发展木质结构材。在乡村、城郊和风景区等土地资源相对丰富区域，积极推进木结构房屋的建设。大力发展性能优良的木质人造板，积极扩大木材特别是竹材在建筑门窗、墙体材料、建筑模板、集装箱底板等方面的应用。拓展木材产品应用范围，适度倡导以木材产品部分替代化石能源产品。

行动13：加强木材高效循环利用。积极推进木材工业“节能、降耗、减排”和木材资源高效、循环利用，大力发展木材精加工和深加工业。针对不同树种、不同树龄、不同部位的材性差异，采取不同加工利用技术，发挥木材的最大功能。要利用原木和采伐、造材、加工剩余物，采用新技术，生产木质重组材和木基复合材。积极发展木材保护业，加快推进木材防腐和人工林木材改性产业化，实现木材保护产品的标准化和系列化，逐步建立和完善木材保护产品质量检验检测体系，改善木材使用性能，延长木材产品使用寿命。要根据《林业产业政策要点》，抓紧制定木材工业实施细则。对限制发展的项目要提出行业准入的具体要求和限制条件。对淘汰的项目要采取得力措施限期淘汰。抓紧制定木材加工业资源综合利用条例、木材综合利用国家标准、木材工业节能降耗标准等一系列促进木材高效循环利用的法律法规和标准。积极推进木材工业企业清洁生产、资源循环利用。加强清洁生产、产品质量和环境认证工作，加强木材高效循环利用监督和产品标识管理，建立绿色环境标志和市场准入制度，从源头上抑制高能耗、高污染、低效益产品的生产。

领域六：湿地恢复、保护和利用

行动 14：开展重要湿地的抢救性保护与恢复。重点解决重要湿地的生态补水问题，有计划地开展湿地污染物控制工作，实施湿地退耕(养)还泽(滩)项目，扩大湿地面积，提高湿地生态系统质量。根据湿地类型、退化原因和程度等情况，因地制宜地开展湿地植被恢复工作，提高湿地碳储量。

行动 15：开展农牧渔业可持续利用示范。建立国家级农牧渔业综合利用示范区、农牧渔业湿地管护区、南方人工湿地高效生态农业模式示范区、红树林湿地合理利用示范基地，优化滨海湿地养殖，实施生态养殖，促进中国农牧渔业对湿地的可持续利用，减少湿地破坏导致的温室气体排放。

二、林业适应气候变化的重点领域和主要行动

领域一：森林生态系统

行动 1：提高人工林生态系统的适应性。尊重自然规律和经济规律，根据未来气候变化情景，从增强人工林生态系统的适应性和稳定性角度，科学规划和确定全国造林区域，合理选择和配置造林

树种和林种，注意选择优良乡土树种和耐火树种，积极营造多树种混交林和针阔混交林，构建适应性和抗逆性强的人工林生态系统。同时，在造林过程中，要把营造林技术措施和森林防火有机结合起来，减少森林火灾隐患。要充分考虑林分的长期和短期固碳效果，科学选择喜光和耐荫树种，尽可能形成复层异龄林。在干旱和半干旱地区，稳步推进防沙治沙工作，因地制宜地加大人工造林种草、封山育林育草措施，合理调配生态用水，建立和巩固以林草为主体的生态防护体系。加强人工林经营管理，提高人工林生态系统的整体功能，保护生物多样性。进一步扩大生物措施治理水土流失的范围，减少水蚀、风蚀导致的土壤有机碳损失。

行动2：建立典型森林物种自然保护区。要在现有自然保护区基础上，进一步针对分布在不同气候带的面积较小且分布区域狭窄的森林生态系统类型，以及没有自然保护区保护或保护比例较少的森林生态系统类型，建立典型森林物种自然保护区，尽快将极度濒危、单一种群的陆生野生物种及栖息地纳入自然保护区，优先保护种群数量相对较少、分布范围狭窄、栖息地分割严重的陆生野生动物。按照统一规划、统一管理和按行政区域分块的办法，对属于一个生物地理单元、生态系统类型相同或相近的自然保护区进行系统整合，构建完整的保护网络，保证生态系统功能的完整性，提高自然保护区的保护效率。

行动3：加大重点物种保护的力度。一是对于亟待保护的重点物种，要根据其特点和空缺程度不同，分轻重缓急，采取就地保护措施。二是对于需实行抢救性就地保护的物种，要对尚未纳入自然保护区网络的栖息地或原生地，优先划建自然保护区；没有条件建立规范性保护区的地段，划建保护小区或保护点，由相邻的国家级自然保护区管理；对大部分种群还没有纳入保护区网络的物种，要适度扩大已有自然保护区规模，使全部或大部分种群及栖息地得到保护；对一些以保护重点物种为主的地方级自然保护区，在具备一定规模和条件时，可升级为国家级自然保护区。对国家级自然保护区内的栖息地较小的种群，应努力改善和扩大种群栖息地。三是对于需要重点进行就地保护的物种，针对其集中分布的物种，应选择几个国家级自然保护区作为核心保护区。在保护空缺处，根据建设条件，划建新的保护区，扩大受保护种群及栖息地的比例。对迁徙性或活动范围较大的野生动物，在主要栖息地和活动通道上建立自然保护区群，注重保护区之间的连通。四是对生境依赖性强的物种，应加大对所处生态系统的保护来保护、恢复和扩大物种种群及其栖息地。五是对广域分布的物种，有针对性地选择条件较好的自然保护区加以重点建设，提高自然保护区水平。六是对已分布在国家级自然保护区内的物种，应加大对自然保护区建设的投入力度，改善、恢复和扩大栖息地。

行动4：提高野生动物疫源疫病监测预警能力。坚持"加强领导、密切配合，依靠科学、依法防治，群防群控、果断处置"的方针，做好野生动物疫源疫病监测防控工作。进一步加强和完善监测体系建设和应急管理，做好野生动物疫病本底调查。全面加强野生动物疫源疫病监测预警工作和国家级监测站建设和管理。加强与卫生、农业等部门的合作，形成联防联动机制。加强人员培训，做好应急演练。

领域二：荒漠生态系统

行动5：加强荒漠化地区的植被保护。针对分布在大江大河源头，且遭受人为破坏严重的半乔木、灌木、半灌木和垫状小半灌木荒漠生态系统，建立一批自然保护区。在保护西部地区独特的植被资源、提高其适应气候变化的能力的同时，增强这类生态系统碳吸收功能，改善西部地区脆弱的生态环境。保护区发展重点是将还没有建立国家级自然保护区的荒漠植被类型，特别是一些面积较小、分布区域狭窄的荒漠植被类型，全部纳入自然保护区，如淡枝沙拐枣荒漠等。

领域三：湿地生态系统

行动6：加强湿地保护的基础工作。开展第二次全国湿地资源调查，全面掌握中国湿地资源的生态特征、社会经济状况、面临的主要威胁和发展趋势，对湿地生态系统储碳量和固碳能力进行调查评估。尽快开展第二次全国泥炭资源调查，摸清中国泥炭资源的分布、储量、保护和开发利用现状及变化趋势等。

行动7：建立和完善湿地自然保护区网络。加强现有湿地自然保护区建设，按照《全国湿地保护工程规划》及《全国湿地保护工程实施规划(2005～2010)》的要求，完善湿地保护基础设施建设，建立健全保护区管理机构，开展社区共管。重点加强对滨海湿地、沼泽湿地、泥炭等湿地类型的保护，发展湿地公园，逐步遏制湿地面积萎缩和功能退化的趋势，形成湿地自然保护网络和较为完整的湿地保护与管理体系。

第六部分

保障措施

一、加强领导，积极开展林业应对气候变化行动

各级林业部门要提高对林业在应对气候变化中特殊地位和作用的认识，将林业工作和应对气候变化工作有机结合起来。要切实加强组织领导，认真履行职责，积极采取措施，结合本地区实际，认真贯彻落实林业应对气候变化的各项措施和目标，真正发挥林业在减缓与适应气候变化中的作用。

二、强化科技，推进林业应对气候变化科学研究

要针对国家应对气候变化战略确定的林业任务和主要行动开展相关研究。一是要紧跟国际研究前沿，深入开展森林对气候变化响应的基础研究，深入探讨气候变化情景下的森林、湿地和荒漠生态系统的碳、氮、水循环过程和耦合机制；开展人类活动对森林、湿地和荒漠等生态系统碳源/汇功能的影响机制研究。二是要结合中国森林的地理分布区域和生态环境类型特点，加强森林生态系统定位站的规划和建设，强化森林生态系统对气候变化响应的定位观测。通过开展生物多样性、森林火灾和森林病虫害等定位观测技术研究，逐步完善森林生态系统观测网络和监测体系，并在此基础

上，加强森林适应气候变化的政策建议、技术选择、成本效益以及适应效果评价等研究，不断提高林业适应气候变化的能力。三是要加强林业碳汇计量和监测体系研究，尽早建立国家森林碳汇计量和监测体系，实现相关数据共享。四是要加强林业减排增汇技术研究。针对林业生物质能源林高效培育、生物柴油提取、木纤维转化乙醇、生物质发电等方面开展合作研究；继续开展重点工程和区域减排增汇潜力与成本效益分析，从碳汇能力、木材供应、水源涵养、生境保护等角度，继续研究相关林种树种搭配模式、多重效益兼顾的栽培和采伐方式、湿地与红树林等生态系统恢复和重建技术、可持续经营技术和农林复合系统的经营技术等。五是加强林业防灾体系研究。要继续研究气候变化情景下的森林火灾和森林病虫害发生机理研究，加大火灾防控新技术、新装备的研发引进力度；提出主要林业有害生物灾害影响和防控对策；加强气候变化下各类野生动物疫病，特别是人兽共患病致病机理的研究，逐步掌握野生动物疫病发病机理和快速诊断、检测关键技术。六是加强气候变化情景下森林、湿地、荒漠、城市绿地等生态系统的适应性问题，提出适应技术对策，开展相关的适应成本和效益分析与评价等。

三、注重培训，提高林业从业人员的工作能力

一是强化应对气候变化相关的专家队伍建设，通过科研立项、建立公平竞争机制等措施，积极鼓励中青年科技工作者积极从事林业应对气候变化相关领域科研，逐步培养一批政治素质高、科研能力强、工作作风实的气候变化专家。二是加强林业应对气候变化相关人员培训。此类培训要和贯彻实施《全国林业人才工作“十一五”及中长期规划》、《全国林业从业人员科学素质行动计划纲要》、“林业专业技术人才知识更新工程”等紧密结合起来。在林业从业人员中树立可持续发展、节约资源、保护生态、改善环境、合理消费、循环经济等观念。三是要积极组织编写《林业应对气候变化地方领导培训大纲》，将生态系统与气候变化关系、地方政府在加强生态

建设与保护、应对气候变化过程中的责任与作用、林业应对气候变化措施等作为开展地方党政领导干部林业专题培训的重要内容，开发相关培训课程和教材。四是积极举办林业应对气候变化专题研讨班和讲座，并将相关内容纳入各类干部职工培训计划。

四、深入宣传，不断提高公众应对气候变化意识

一是广泛开展科普宣传，深化全社会对林业的功能与作用的认识。宣传普及森林培育经营管理知识、森林和湿地的吸碳、贮碳的增汇功能和作用，让公众认识到林业是经济有效的减缓气候变暖的重要措施，在应对气候变化中具有不可替代的重大作用，促进全社会重新审视林业的地位和作用，增强公众的生态意识和保护气候意识，动员更多的人参与到“造林增汇、保林固碳、改善环境、应对气候变化”的行动中。二是搞好动态宣传。积极宣传林业在应对气候变化方面的重要举措和具体行动，以及这些举措在增加森林碳汇、防止水土流失、治理荒漠化、消除贫困、保护生物多样性等方面取得的多重效益。通过多种媒体和手段，促进人们进一步“关注森林”。使林业应对气候变化宣传实现全方位和多视角。三是加强典型宣传。突出宣传各地生态建设的典型，报道重点地区在推进植树造林、改善生态状况等方面取得的变化和经验。结合已成立的中国绿色碳基金，大力宣传各级社会团体、组织和重点企业参与造林绿化、减排增汇的行动。选择并树立一批以实际行动参加林业应对气候变化行动的企业、团体、组织和个人的典型事迹，通过典型带动全社会的行动。

五、创新机制，推进林业改革和应对气候变化工作

一是要贯彻落实《中共中央 国务院关于全面推进集体林权制度改革的意见》，明晰产权、完善政策，给予农民更多的发展权利，增强农民的发展能力，保障农民的合法权益，进一步调动广大林农造林护林和经营森林的积极性。二是要大力支持各类社会主体参与

林业建设，以政策为引导，以法律作保障，鼓励和扶持社团、企业、外商等开展多种形式的造林，不断扩大非公有制造林的数量和规模。三是要制(修)订林业保护法律法规，完善相关配套政策，用法律手段和政策保障促进林业发展。适时修订森林法，争取尽快出台国家湿地保护条例，加大执法力度，扩大社会监督。四是要继续完善各级政府造林绿化目标管理责任制和部门绿化责任制，进一步探索市场经济条件下全民义务植树的多种形式，推动义务植树和部门绿化工作的深入发展。五是要充分发挥中国绿色碳基金平台作用，积极鼓励企业、组织、团体、个人以自愿方式加入中国绿色碳基金。对碳汇计量、监测、注册、登记等工作进行资质管理，逐步建立资质认证制度。

六、突出重点，增加林业建设资金

一是要在公共财政体制下，保持对林业发展和保护工作的持续资金投入。要根据林业应对气候变化的重点领域和主要行动，继续保持和加大公共财政对重点工程造林、森林可持续经营、珍贵树种营造、森林火灾和病虫害预防、野生动物疫源疫病监测防控、森林保护等方面的资金支持力度，确保林业应对气候变化的重点领域和主要行动得到有效实施。二是要多渠道筹集林业发展和保护的资金，积极组织编制林业利用外资的项目规划，引导外资流向林业重点工程，进一步完善国家林业信贷资金投入政策和管理体制。三是支持木材高效循环利用，积极争取国家扶持企业在提高木材综合利用率方面的技术改造。四是要加大对林业应对气候变化所涉及的森林碳汇计量和监测、森林恢复技术、困难立地造林技术、可持续经营综合技术、森林适应性评估等方面的专项科研以及与林业应对气候变化相关的能力建设、宣传培训、国际履约等方面的经费支持。

七、服务大局，积极开展林业国际合作

一是以“亚太森林恢复与可持续管理网络”为平台，积极推进区

域性林业国际合作。二是要全面深入地参与《公约》和《京都议定书》国际进程中林业议题谈判活动，积极组织部门内外专家针对林业议题开展谈判对策研究；积极支持中国林业专家参与 IPCC 及相关工作。三是要积极推进开展清洁发展机制下碳汇造林活动，积累项目实施经验，探索借鉴国际机制推进国内造林工作的途径。四是要鼓励开展林业与气候变化相关的双边和多边合作以及对话机制，利用国际合作资源提高推进林业应对气候变化能力建设，积极促进发达国家先进林业经营理念和经营技术转让。积极争取在援外渠道中，增加与发展中国家在林业应对气候变化领域的合作。

The Forestry Action Plan to Address Climate Change

State Forestry Administration, P. R. China

November 6, 2009

Contents

Part 1

Introduction

Since the 1970s, the global climate change① featured by global warming has drawn more and more attention from the international community and become a hot issue in the domains of international politics, economy, environment and foreign relationship.

In 2007, the Intergovernmental Panel on Climate Change② (hereinafter referred to as IPCC) officially released the Fourth Assessment Report, in which a large amount of statistics once again proved that the global climate change was an unequivocal fact. In the next 100 years, global warming will continue and have a great impact on the ecosystem and the

① According to the definition of the *United Nations Framework Convention on Climate Change*, "Climate change" means a change of climate which is attributed directly or indirectly to human activity that alters the composition of the global atmosphere and which is in addition to natural climate variability observed over comparable time periods.

② Intergovernmental Panel on Climate Change (IPCC) is a governmental organization established jointly by World Meteorological Organization (WMO) and the United Nations Environment Program (UNEP) in 1988. It aims to assess the global climate change comprehensively, objectively, openly and transparently.

human's survival. Although global warming is caused by both natural and human factors, it is mainly the result of the increase of the density of greenhouse gases① including the carbon dioxide released into the atmosphere due to human's overuse of fossil energy and deforestation for land reclamation since the industrialization. This report showed that the carbon dioxide in the atmosphere increased to 379ppm② in 2005 from 280ppm before the industrialization, leading to an increase of 0.74℃ in global temperature in the past 100 years, causing the rise of sea levels, the melting of ice and snow on mountains, an obvious change in the distribution, frequency and intensity of precipitation, and an increase of extreme weather events, which posed severe threats to the sustainable development of global ecosystem and human society. If no effective measures were to be taken to control the greenhouse gas emission, the density of greenhouse gases in the atmosphere would continue to rise, which would lead to an increase of 1.4 ~ 5.8℃ in the average global temperature by the year of 2100 and consequently result in an irreversible impact on the global ecosystem and the survival and development of human society③.

The global warming is having obvious impacts on China as well. The

---

① All gases which can cause the increasing of atmosphere temperature are referred to as Green House Gases (GHG). There are many different kind of GHG, according to UNFCCC, the following gases are included as GHG to be controlled: CO_2, CH_4, N_2O, HFCs, PFCs, and SF_6, of which, CO_2 is regarded as the main GHG.

② Parts per million.

③ Please refer to the related contents of the Fourth Assessment Report issued by IPCC.

meteorological observation statistics show that, in the past nearly 100 years, China's average land surface temperature has increased by 0.5 ~ 0.8℃. In the past 50 years in particular, China's average land surface temperature has increased by about 1.1℃, averaging an increase of 0.22℃ per decade, apparently higher than that of the world or northern hemisphere over the same period of time. Since the 1950s, the sea level has an annual increase of 1.4 ~ 3.2mm in the coastal regions of China, the ice coverage in the Bohai Sea and the north of the Yellow Sea has decreased, the area of glacier in northwest has dropped by 21%, the depth of frozen soil in Tibet has decreased by 4 ~ 5 meters, the water level of some inland lakes in plateau areas has increased, and the grass production of the pastures in Qinghai province and the south Gansu province has decreased. Since the 1980s, China's spring phonological period has advanced by 2 ~ 4 days, the drought-hit areas in the north has expanded, the flood in the south has become more serious, and there has even arisen the phenomenon of coral bleaching① in such coastal areas as Hainan and Guangxi.

In order to safeguard the global ecological security and the sustainable economic and social development of the human beings, the global warming should be addressed by mitigation and adaptation. Mitigation means reduction of the greenhouse gas emission by improving energy efficiency or reducing energy consumption in industry or energy production process, or by reducing the density of greenhouse gases in the atmosphere through developing and protecting forest to enhance the absorption of greenhouse gases. Adaptation means to take proactive

---

① Please refer to *National Assessment Report on Climate Change*, Science Press

measures to strengthen the adaptability of both the natural ecosystem and human beings to the global warming and to prevent or reduce the negative impacts of global warming.

As a developing country, China does not assume the task of quantifiable greenhouse gases emission reduction, but the Chinese government has fully realized that it is a common responsibility of the whole mankind to address global climate change and safeguard the security of global ecosystem, and that to control and reduce the greenhouse gas emission is not only in line with the long-term development goal of establishing a resource conservation, environment-friendly, low-energy-consumption and low-emission society, but also an internal requirement to carry out the scientific outlook on development and achieve the sustainable economic and social development.

The Chinese government has been attaching great importance to the issue of climate change over the past few years. As early as February 1990, the State Council established the National Coordination Group to Address Climate Change (NCGACC), responsible for coordinating and formulating policies and measures related to climate change. The daily work of the NCGACC was undertaken by China Meteorological Administration. After 1998, the State Council restructured the NCGACC, the responsibility of the daily work of the NCGACC was shifted to the National Development and Reform Commission (NDRC), and the main tasks of the NCGACC were to formulate key national strategies, principles and policies on climate change, coordinate and solve important issues related to climate change. In 2007, the NCGCC was further restructured and the National Leading Group on Addressing Climate Change, Energy Saving and Emission Reduction headed by

Premier Wen Jiabao was established, the daily work of the Leading Group was still undertaken by the NDRC. Currently, 20 national departments and institutions are involved in addressing climate change.

In order to practically implement China's obligations under the *United Nations Framework Convention on Climate Change* (UNFCCC) and expound China's policies and views on addressing climate change to the international community, the State Council issued *China's National Climate Change Program* (CNCCP) in June 2007. The CNCCP put forward forestry measures to address climate change, including increasing forest sequestration of greenhouse gases, maintaining and expanding the comprehensive functions of forest ecosystem, and building sound ecological environment. Soon after the issuance of the CNCCP, the Meeting of the National Leading Group on Addressing Climate Change, Energy Saving and Emission Reduction was held to make overall arrangements for carrying out the CNCCP, and impose specific requirements on relevant departments and local governments for the implementation of the CNCCP. In 2007, President Hu Jintao made an important proposal at the 15th Asia-Pacific Economic Cooperation (APEC) forum to establish an Asia-Pacific Network on Forest Rehabilitation and Sustainable Management, which was praised by the international community as the Forestry Program to address climate change. State Forestry Administration responded proactively to this proposal, and set up a Leading Group on Addressing Climate Change, Energy Saving and Emission Reduction and its office within the SFA in July 2007, and started to organize the formulation of *the Forestry Action Plan to Address Climate Change* (hereinafter referred to as *the Forestry Action Plan*) to carry out the tasks imposed by the CNCCP to forestry, and provide guidance to forestry departments at all levels in undertaking relevant work to address climate change.

Part 2

Forestry and Climate Change

1 Forests' Unique Role in Addressing Global Climate Change

Forest is the biggest carbon sink and the most economical carbon absorber on the land. According to the IPCC estimates: about 2.48 trillion tons of carbon is stored in global terrestrial ecosystems, among which 1.15 trillion tons of carbon is stored in forest ecosystems. Through photosynthesis, forests absorb carbon dioxide and give off oxygen, store carbon dioxide from the atmosphere in vegetations and soil, and the process is called carbon sink. The scientific research indicates: as the trees grow 1 m^3, they will absorb 1.83 tons of carbon dioxide and give off 1.62 tons of oxygen in average. The carbon sequestration and stores in global forests account for 90% of the annual carbon flow between the atmosphere and the earth's surface. Like their other important ecological functions, the carbon sink function of forest plays a significant role in maintaining global ecological security and climate security.

The sharp forest reduction is a main factor of global climate change. The global warming mainly results from the greenhouse effect caused by the rise of CO_2 and other greenhouse gas concentrations in the atmosphere. The rise of CO_2 concentration in the atmosphere mainly has two reasons: firstly, CO_2 emission from the large scale of burning fossil fuel; secondly, CO_2 release from the sharp reduction of global forests. The area of global forests has been reduced from 7.6 billion ha in the beginning of human civilization to the current 3.8 billion ha. *Global Environment Outlook* 2000 of the United Nations indicates that global forests have been reduced by 50%. At present, the reducing trend of forest still maintains. The report of Food & Agriculture Organization of the United Nations (FAO) showed: from 2000 to 2005, the global annual average deforestation① area was 7.3 million ha.

Forest rehabilitation and protection is one of the most fundamental measures to mitigate global climate change. IPCC indicated in its fourth global climate change assessment report issued in 2007: forestry measures could highly reduce greenhouse gas emission and increase carbon sink at quite low cost, so as to mitigate climate change. Centering on the second-commitment-period negotiation under the *Kyoto Protocol*, many countries and international organizations actively advocate to mitigate climate change by the rehabilitation and development of forest ecosystems.

Meanwhile, climate change seriously affects forests. Forest growth is affected by natural factors including sunlight, temperature, moisture and wind all the times, while these factors are closely linked with climate.

---

① Deforestation referred in the Convention indicates the situation that converting forest land to non forestry land, such as converting forest land to farmland, grazing land, or urban infrastructure construction land, etc.

Therefore, we should positively improve forests' ability to adapt to climate change, reduce climate change's negative impact on forests, and maintain sound ecological functions of the forest.

2 China's Achievement in Forestry Development and Contributions to Global Climate Change Mitigation

The Chinese government has long attached great importance to forest development and protection. Since 1978, the Chinese government has successively implemented a series of regional shelterbelt development programs in the Three North (Northeast, Northwest and North) area, coastal area, plain area, the upper and middle reaches of the Yangtze River, the Taihang Mountain, the vicinity area of Beijing and Tianjin, the Huaihe River and the Taihu Lake valley, the Pearl River valley, the Liaohe River valley, etc. Since the forestry development layout adjustment in 1998, the Chinese government has launched pilot sites and successively implemented six national key programs, including the Natural Forest Protection Program, the Conversion of Cropland to Forests Program, the Sandification Control Program for the Areas in the Vicinity of Beijing and Tianjin, Key Shelterbelt Development Program in Such Regions as the Three North and the Middle and Lower Reaches of the Yangtze River, the Forest Industrial Base Development Program in Key Regions with a Focus on Fast-growing and High-yield Timber Plantations, and the Wildlife Conservation and Nature Reserve Development Program. By 2008, six national key programs had accomplished 51.5374 million ha of plantations (including 14.7538 million ha of mountain closure). The total investment was RMB 278.126 billion, among which the national investment was RMB 241.636 billion①. Since 1981, China has conducted the Nationwide Voluntary Tree-Planting Campaign. By the end of 2008, 11.52 billion person-times had been involved in the voluntary tree planting activities with a total of 53.85 billion trees established, the urban green cover increased from 10.1% in 1981 to 35.29%, the public green land area per capita increased from 3.45 m^2 to 8.98 m^2, the urban and rural greening was promoted and the human living environment was improved②.

In order to protect the forest, China has successively issued 9 forestry laws, 15 forestry administrative legislations, 43 forestry regulations, and over 300 local legislations and regulations, and has established a forest resource conservation law system with *Forest Law*, *Wildlife Conservation Law*, and *Law on Prevention and Control of Desertification* as the core, and a forest resource management system with the main content of forest administration and supplementary content of resource monitoring and supervision. China has implemented several actions against illegal crimes like excessive logging, unauthorized use and expropriation of forest land and wetland, etc. China handled 3.317 million cases concerning the destruction of forest resources from 2001 to 2008. Meanwhile, China has strengthened control and prevention of forest fires and pests and diseases, and development of nature reserves. Currently, China has established 2531 nature reserves of various types,

---

① *China Forestry Statistics Yearbook* (2008), State Forestry Administration, China Forestry Publishing House.

② *China National Land Greening Situation Bulletin for* 2008, published by National Greening Commission Office in March 11, 2009.

accounting for 15.2% of the total land area①.

By adopting a series of measures to develop and protect forest resources, China continuously increased its forest area and stock volume. According to the 6th national forest inventory (1999 ~ 2003): China's forest area had reached 175 million ha, the forest cover was 18.21%, accounting for 4.5% of the global forest area, ranking the 5th in the world; the forest stock volume was 12.456 billion m^3, accounting for 3.2% of the total volume of the world, ranking the 6th in the world; the preserved plantation forest area was 54 million ha, accounting for 1/3 of the global plantation forest area, ranking the 1st in the world②.

China's achievement in forestry development has been widely recognized by the international community. According to the FAO's *State of the World's Forests* 2005: from 2000 to 2005 global forest resources maintained a decrease, while net increase appeared in forest area in the Asia-pacific region. The increase of China's forest resources greatly neutralized the high logging rate in other regions③. FAO's *State of the World's Forests 2009* again affirmed China's achievements in the continuous increase in forest resources④.

While the sustained increase in China's forest area and stock volume

---

① Reference from relevant articles in *China Forestry Development Report* (2001 – 2008), State Forestry Administration, China Forestry Publishing House.

② *China Forest Resource*, Chief Editor Lei Jiafu, China Forestry Publishing House.

③ *The State of the World's Forests* (2005), FAO.

④ *The State of the World's Forests* (2009), FAO.

improved China's timber supply and ecological environment, it also absorbed and fixed large volume of CO_2. It is estimated that from 1980 to 2005, China accomplished a net absorption of 4.68 billion tons of CO_2 with continuous afforestation and forest management activities, and reduced 430 million tons of CO_2 emission by controlling deforestation, totaling 5.11 billion tons①. China has made significant contributions to mitigate global warming.

The carbon storage in the forest reflects the general situation of forest ecosystem's absorption and fixation of CO_2. Different calculation methods may cause obviously diverse calculation results. Based on the data of national forest inventory (1977 ~ 2003), the academicians including Fang Jingyun analyzed and concluded that since the end of 1970's, China's carbon sink in forest vegetations had reflected a distinct increasing trend, the forest carbon density per unit area had increased from 36.9 tons of carbon per hectare in the early 1980s to 41 tons of carbon in 2003.

Although the research results of the future trend of China's forest carbon sink differ by different research methods, the general trend is: from 1990 to 2050, the carbon storage in China's forests will gradually increase.

3 Initial Analysis of China's Methods and Potentials in Mitigating Climate Change

Forestry's role in mitigating climate change is mainly reflected in 4 ways, namely, increase of carbon sink, reduction of emission, storage and alternative fuels. Concrete measures include increasing forest carbon

① *China's National Climate Change Program.*

absorption through afforestation, vegetation rehabilitation, sustainable forest management activities; reducing carbon emission from the forest by appropriate logging control, reducing deforestation, prevention and control of forest fires, pests and diseases; expanding the carbon storage in woody forest products by extending the utilization of woody forest products and prolongating the timber use duration; converting woody forest products and wood nuts into biofuel to partly replace fossil fuels, such as forest logging and processing residuals energy utilization, converting the wood nuts into bio-diesel, etc, which will help reduce the fossil fuel utilization and carbon emission. China has quite great potentials in developing forestry biomass energy, which should be actively developed and utilized.

3.1 Increase carbon sink through expanding forest area by afforestation. China's forest coverage is comparably lower than the major developed countries and some of the developing countries. There are still 57 million ha of barren hill and land, 54 million ha of sandy land suitable for plantation, and vast areas of slope cropland above 25 degrees or available land to be utilized for afforestation. At same time, there's good space to increase forest coverage by increasing the utilization rate of existing forest land and developing farmland shelterbelt network. According to the *Resolution on Accelerating Forestry Development by the CPC Central Committee and State Council*, by the year 2050, China's forest coverage will be increased to 26% from the current 18.21%. By then, the volume of forest carbon sink will have been considerably elevated.

3.2 Increase carbon sink through quality improvement of existing forests. The average standing volume of existing forests is only about 84m^3 per hectare, and the average annual growth is about 3.55m^3 per hectare, most forests are plantations or secondary forests with low biomass density. According to the analysis of experts, the carbon sink of China's existing forest vegetation resources is only 44.3% of its potential. Therefore, it's completely possible to greatly increase the carbon sink capacity of existing forest vegetation resources by doubling the average stands growth through adjusting stands structure and intensifying forest management.

3.3 Reduce carbon emission through strengthening forest protection. Firstly, reduce the carbon emission from forest itself, by strict control of arbitrary expropriation of forest land and deforestation. The past forest inventory results indicate that the average annual loss of forested land by arbitrary expropriation of forest land is around 1 million ha. Therefore the strict management of arbitrary expropriation of forest land and deforestation has big potentials to reduce carbon emission. At the same time, in logging operation, scientific planning and low-intensity operation to protect ground vegetation and soil will reduce the carbon emission from the damage of ground vegetation and soil. Secondly, forest fires, pests and diseases can release the carbon stored in the forest ecosystem into the atmosphere in a very short time. Therefore, through strengthening the supervision of the combustibles in the forest, and establishing the forest fire, pest and disease early-warning system, it is possible to effectively control the frequency and the scope of the forest fire, pests and diseases and reduce the forest carbon release.

3.4 Reduce greenhouse gas emission through wetland protection and soil and water erosion control of forest land. Firstly, the wetland soil reserves a lot of organic carbon, which will be decomposed if damaged and greenhouse gases such as CO_2 will be released to the atmosphere. The existing various wetlands in China with an area above 100 ha total 38.48 million ha. A lot of wetlands have degraded or have been expropriated with the economic and social development. Enhancing wetland protection will reduce the greenhouse gas emission caused by damage of wetlands. Secondly, the soil of forests also stores a great deal of organic carbon, accounting for more than 60% of the total carbon storage in the entire forest eco-system. The biological measures which aim to control forest land erosion will help forest soil protection, improve and accelerate forest soil development, and spur non-forest land conversion to forest land. Thus the forest land's capacity of carbon sink can be upgraded.

3.5 Reduce the carbon emission through substitution of fossil fuel with development of forest biomass energy. The direct burning of forest biomass materials, woody fibre conversion to ethanol through hydrolization, etherealization of woody materials through pyrogenation, and obtaining bio-diesel from fruits of oil plants are various ways to partially substitute the fossil energy, and reduce greenhouse gas emission. According to the statistics, in China, the annual residues from logging and timber processing which can be used for energy generating amount to 300 million tons. If all of them can be utilized, about 200 million tons of standard coal will be substituted. Meanwhile, if the existing barren hills and land, saline-alkali land and reclaimed mine land which are difficult to manage can be utilized for development of certain energy plantations, the percentage of forest biomass energy to substitute fossil energy will be increased, which is helpful to reduce China's gross volume of greenhouse gas emission.

3.6 Increase carbon store in woody forest products through extending timber utilization and its service life. The volume of greenhouse gas emission from energy consumption in timber producing and processing is far lower than in steel and aluminum manufacturing. If timber can partly substitute some of the energy-intensive materials, not only can the carbon sink be increased, but also the carbon emission from fossil fuel consumption for material manufacturing can be reduced. A research result shows that by using $1m^3$ of timber to substitute equivalent volume of cement or brick, the CO_2 emission can be reduced by 0.8 ton. It not only saves the energy, but also reduces pollution. As long as the wood product does not decay or burn, it's always an important carbon storage. According to the estimates by experts, the carbon store in woodwork in China amounted to 1.2 billion to 1.8 billion tons of CO_2 equivalents from 1961 to 2004, which is an important contribution of forestry to mitigation of climate change.

4 Influences of Climate Change to China's Forestry

Development of forestry is helpful to climate change mitigation; the climate change will show movement in its factors including temperature, humidity, growing season, rainfall and evaporation, etc. The extreme weather condition will happen more frequently because of climate change and produce a current and potential impact on forestry development. According to *National Assessment Report on Climate Change*, the main impact of climate change on China's forests and forestry development is that the sustained global warming will bring negative impacts to our forest eco-system's stability, structures and functions.

For the vegetation distribution, the climate change may cause a northward movement of vegetations in sub-tropical and temperate zone of east China, advance the phenological period, also move main plantation species northward, and threaten biodiversity. For the productivity of forests, though climate warming may increase the forest productivity in different extents, it will not alter the geographic distribution of the current forest productivity. The increment of forest productivity in most areas of tropical and sub-tropical zones is only 1% , may reach 10% in the cool temperate zone and subalpine forests in southwest, and can reach 2% ~ 8% in warm-temperate and temperate zone. For the habitats of wildlife, the distribution area of some rare species such as *Taiwania flousiana*, *Davidia involucrata*, and the habitats of Giant Panda, Yunnan snub nosed monkey and Tibetan antelope will decrease, some species will go extinct faster due to weak adaptability. For forest disasters, the climate change will lead to abnormal change of regional climate features and routine, the frequency and intensity of forest fires will be uplifted. For example, the number of lightning will increase, the fire-prevention period will be extended, and the extreme fire risks and severity will upgrade. These will directly harm the forest growth, and may damage the forest eco-system structures and functions. Simultaneously, climate change will expand the forest pest and disease distribution zone to the north, advance their occurrence, shorten the period of occurrence, increase their generations, enlarge their scope of occurrence and deepen the damage extent. The climate change also will aggravate the harm of invaded invasive pests and diseases, and affect the pathogens' survival and variation, the vector insects' multiplication distribution and their epidemiologic features, move the distribution latitude of bacteria carriers and diseases upward, which will be much disadvantageous to wildlife survival and multiplication. For the changes of drought and flooding, with future warming, the western desert and grassland may shrink slightly and be replaced by grass and scrubland. But it will accelerate the frozen earth degradation, glacier retreat and water resource shortage, which will further affect the inland rivers. Extreme drought and sub-humid dry area will sharply increase, the total area of desertified and water and soil-eroded land will expand. For the wetland functions, the climate change will bring cutoff of the northern rivers, shrink the lakes, reduce the reservoirs' water level, uplift the sea level, further decrease the wetland area and functions, and badly damage the mangrove eco-system on coastal line.

The shortage of total forest resources in China imposes a great pressure on forest land, wetland and sandy land protection with the acceleration of industrialization and urbanization. This is a great challenge to afforestation and ecological rehabilitation. Effective measures must be taken to upgrade the forests' adaptability to climate change. The upgrading of forests' adaptability to climate change will also further enhance the capability of forests on mitigating the climate change.

Part 3

International Processes in Addressing Climate Change and Forestry

1 Forest-related Issues in International Processes in Addressing Climate Change

The First United Nations Conference on Environment and Development (UNCED) held in Brazil in 1992 adopted the *United Nations Framework Convention on Climate Change* (UNFCCC) and identified the goal to "stabilize greenhouse gas concentrations in the atmosphere at a level that would prevent dangerous anthropogenic interference with the climate system". UNFCCC entered into force in 1994 and China is one of the contracting parties.

To meet the goal of UNFCCC, all country parties should fulfill the obligations stated in the convention, actively adopt mitigation and adaptation measures and increasingly improve their ability to address climate change. Since most of the greenhouse gasses in the atmosphere were emitted by developed countries and according to the principle of "common but differentiated responsibilities", developed country parties should take the lead in reducing emissions. In December of 1997, the *Kyoto Protocol* was adopted at the 3rd Conference of the Parties to the UNFCCC held in Japan. It for the first time legally set out the mandatory emission limits for Annex I countries (including industrialized countries and economies in transition, together called "developed countries"), who should reduce 5.2% on the average of their greenhouse gas emissions below their 1990 levels from 2008 to 2012.

In order to help Annex I countries fulfill their emission reduction commitments under the *Kyoto Protocol*, it offers them three flexible mechanisms. They are emissions trading, joint implementation and clean development mechanism (CDM). Amongst them, virtually, emissions trading and joint implementation are the trading of emission rights of greenhouse gasses, while CDM means that developed countries may earn certified emission reduction (CER) credits through cooperating with developing countries in emission reduction or carbon sequestration increase projects to offset their emission reduction targets prescribed in the *Kyoto Protocol*. In fact, CDM means developed countries buy the emission rights of greenhouse gases from developing countries and it requires developed countries to implement projects in developing countries in the form of capital investment, technology transfer, etc. and to facilitate the sustainable economic and social development in developing countries while obtaining emission rights of greenhouse gases from them.

Carbon sequestration and emission reduction by forest-related activities have been viewed as an important measure to implement UNFCCC and the *Kyoto Protocol*. All parties to the UNFCCC have held prolonged negotiations on how to help developed countries meet their emission targets by forest-related activities since the adoption of the protocol and worked out a series of decisions by the party conferences. In one word, there are two ways: one is that developed countries may respectively use their own carbon sequestrations produced by forest-related activities since 1990 to offset their greenhouse gas emissions from 2008 to 2012; the other is that developed countries may use CDM to purchase carbon credits produced by afforestation/reforestation projects implemented in developing countries to partially offset their greenhouse gas emissions from 2008 to 2012. According to related decisions made by party conferences, developed countries could fulfill around 20% ~ 30% of their emission-reduction targets prescribed in the *Kyoto Protocol* by using forest carbon sequestration. Thanks to the low costs of forest carbon sequestration, the pressures on developed countries to fulfill their commitments under the protocol can be relieved to a great extent.

Meanwhile, some tropical developing countries have long been challenged by serious deforestation. The IPCC Assessment Report shows that the CO_2 emitted by deforestation all around the world is more than that produced by the transportation sector and it is the third largest source of emission behind energy and industry, taking up around 20% of the total greenhouse gas emissions in the world. Following a series of negotiations, the 13th Conference of Parties to the UNFCCC held at the end of 2007 in Bali, Indonesia included reduction of emissions from deforestation and forest degradation in developing countries into the *Bali Action Plan*. How to give play to the important role of forestry of developing countries in mitigating climate change has been an important part of the common action to mitigate the global climate change in the future.

Owing to its special role in addressing climate change, in the international processes in addressing climate change, forestry is always shouldering a crucial mission either in helping developed countries fulfill their quantified emission-reduction commitments or in further facilitating the participation of developing countries in actions to mitigate the global climate change. It can be foreseen that this will bring many challenges and opportunities for forestry development.

2 Challenges to Forestry Development Brought by Climate Change

2.1 Climate change will bring great impacts on the forest productivity, species distribution and the stability of ecosystem. In case that the negative impacts of climate change on forests could not be well prevented and controlled, forests would not only be unable to mitigate climate change, but also would accelerate it and in return affect their own healthy development. In recent years, climate change has increased the frequencies and intensity of forest fires, pests and diseases in many places across China and worsened the shortage of water resources in the arid and semi-arid areas in west China. Generally speaking, climate change will make it harder to protect and develop the forest resources in China.

2.2 Climate change will intensify the conflicts between different land types and land uses. Researches show that climate change may have great impacts on the layout and structure of China's agricultural production and cause the decrease of plantation productivity. The increasing population means that more forests or forest land will be deforested or expropriated for food and animal husbandry and the conflicts between different land uses will certainly be more acute. It will make it more difficult for the forestry sector to manage forests and forest land and will restrict the carbon sequestration increases by increasing forest area.

2.3 Climate change has impacts on the supply of timber and non-timber forest products (NTFPs) as well as ecological forest services. Many studies show that although multiple benefits could be brought by using forestry solutions to mitigate climate change and the costs may be reduced, the structure of land uses would be changed. In addressing climate change, it requires to reform and innovate the existing forestry policies, systems and mechanisms in China to balance the demands for forest products and various ecological products including carbon sequestration and to provide continuously effective incentives for local forest managers.

2.4 As the increasingly deepening of UNFCCC negotiations, actions to reduce emission from deforestation and forest degradation in developing countries will be gradually included into the regime of climate change mitigation and the costs of forest logging and utilization will certainly increase. It will to some extent increase the costs for China to import timbers and restrict our utilization of forest resources from overseas. This has proposed new requirements for China to readjust and improve forest-related policies and measures to ameliorate the capability of self-supply of wood.

3 Opportunities to Forestry Development Brought by Addressing Climate Change

3.1 The Fourth IPCC Assessment Report realizes that forestry is an important technically and economically viable and low-cost measure to mitigate climate change at present, in the future 30 years or even longer period. It can produce synergy effects with adaptation and bring multiple benefits including increase of employment and income, protection of water resources and biodiversity and poverty alleviation while mitigating the warming. In the context of climate change, promoting the role of forestry in mitigating climate change will raise the social awareness of the importance of forest values and the work of forestry and produce a sound social environment with great attention to forestry and its development.

3.2 The innovative mechanisms under UNFCCC and the *Kyoto Protocol* provide new opportunities for the development of forestry. In particular, the appearance and development of the carbon market which is based on the right of emission help set a price for the practices of CO_2 emissions. The pricing mechanism can not only restrict the emission practices of the principal emitters, but also cut down the total costs for the reduction of global greenhouse gas emissions. Forest carbon sequestration is a part of the global carbon trading. To internalize the externalities of the functions and benefits of forest carbon sequestration through trading carbon on the carbon market will help integrate systematically the interests of the users and suppliers of forests' ecological benefits and further improve the compensation mechanism for ecological benefits in the near future and push forward the reform and innovation of investment and financing mechanisms for forestry development in the long run.

3.3 Based on the *Bali Action Plan*, it has been an important part for developing countries to participate more in the climate change mitigation actions after 2012 to reduce the CO_2 emissions caused by deforestation and forest degradation in these countries and increase carbon sequestration through forest conservation, sustainable forest management and afforestation. Whether developing countries can take effective actions in this regard depends on to what extent the developed countries will offer financial and technical supports. Therefore the inclusion of forestry in international and national processes in addressing climate change will provide new opportunities for forestry development.

3. 4 Giving full play to forestry in addressing climate change does not only involve afforestation and forest management, but also the development of bio-energy to replace fossil fuels and the utilization of biomass materials to substitute the materials produced with fossil energy. For example, oil extracted from fruits of oil plantations can be turned into biological diesel oil; oriented energy plantations, logging residues and wastes from timber processing can be used for power generation or heating; ethanol fuels produced from semicellulose biomass can be used as the second generation bio-fuel; and wood can directly replace part of fossil energy to produce bricks, steels, aluminum, glass and other materials, which can not only enormously reduce the emission of greenhouse gasses, but also provide a new growth point for forestry development and the sustainable economic and social development at large.

3. 5 China is now actively participating in the afforestation/reforestation projects under CDM as stipulated in the *Kyoto Protocol.* CDM has not only introduced certain afforestation funds to China, but also provided chances for us to master related international rules and undertake carbon inventory, monitoring, verification and trading. It has helped us strengthen our capability to participate in the implementation of carbon sequestration projects and provided references for us to further improve payment for ecological services in virtue of market-based mechanisms, expand funding channels for afforestation/reforestation and accelerate the afforestation/reforestation process in China.

In conclusion, forestry development is faced with both tremendous challenges and strategic opportunities in the context of climate change. Climate change will urge various governments to pay more attention to forestry and speed up the reform of forest administration system and the innovation of forestry development mechanisms. It will bring new driving forces for the forestry development in all countries if we can seize the opportunity on our own initiatives and respond to challenges actively.

Part 4

Guidelines, Basic Principles and Major Objectives for Forestry to Address Climate Change

1 Guidelines

With the guidance of scientific outlook on development, and in accordance with the policies and measures stipulated in the *National Climate Change Program* for forestry to address climate change as well as the middle-and-long-term forestry development plans, we shall accelerate the implementation of key forestry programs to enlarge forest area, improve forest quality, and strengthen the protection of forest eco-system, wetland eco-system, and desert ecosystem. We shall rely on the scientific and technological development to transform increment methods, make overall plans to promote the development of forestry ecological system, industrial system and eco-culture system, continuously enhance the functions of forestry to sequestrate carbon, increase the capabilities of forestry in China to mitigate and adapt to climate change, and make new contributions to developing the modern forestry, building the conservation culture, and pushing the scientific development forward.

2 Basic Principles

2.1 Integrate forestry development objectives with the national strategies to address climate change. To identify forestry development objectives shall take into full consideration the national strategies to address climate change, and integrate the enhancement of the economic, ecological, and social functions of forestry with the improvement of forestry capability to mitigate and adapt to climate change. The formulation of policies and strategies at various levels to address climate change shall attach enough importance and support to forestry as one key measure.

2.2 Integrate increase of forest area with improvement of forest quality. On the one hand, we shall continuously enlarge forest area and accelerate the rehabilitation of degraded wetlands and the management of sandified lands, so as to increase carbon sequestration. On the other hand, we shall make efforts to increase the annual increment of per unit forest area and its capacity to store carbon. By scientifically managing forests, we shall gradually transform forest stands with low biomass and carbon density to those with high biomass and carbon density, and intensify the capacity of existing forests to store carbon and related comprehensive benefits.

2.3 Integrate increase of carbon sequestration with control of carbon emission. While expanding forest area, accelerating the rehabilitation of degraded wetlands and the management of sandified lands, and improving the quality of existing forests, in order to increase carbon sequestration, we shall also adopt active measures to protect the resources of forests, wetlands and deserts ecosystems, in order to protect forest, wetland and desert ecosystems from destruction, and prevent the re-emission of carbon stored in these ecosystems into the atmosphere.

2.4 Integrate the leading role of governments with the social participation. We shall not only bring into play the leading role of governments in the process of the forestry development, but also shall continuously encourage the participation of the whole society and people in forestry development. By various means, we shall motivate enterprises, social groups, organizations, and individuals to actively join in such activities as afforestation, tree-planting, forest protection, increase of carbon sequestration to address climate change.

2.5 Integrate mitigation with adaptation. We shall not only increase carbon sequestration and decrease carbon emission of forests, in order to enhance the functions of forestry to mitigate climate change, but also shall pay great attention to the capacities of forestry to adapt to climate change as the basis for strengthening the functions of forestry to mitigate climate change, so as to develop coordinated effects between mitigation and adaptation of forestry to address climate change.

3 Major Objectives

3.1 Overall objective. The overall objective is to accelerate afforestation efforts at barren hills and lands liable to forests, expand the area of wetland rehabilitation and conservation, and speed up sandy land management; to continuously implement the Natural Forest Protection Program, the Conversion of Croplands to Forests Program, the Sandification Control Program for the Areas in the Vicinity of Beijing and Tianjin, the Forest Industrial Base Development Program in Key Regions with a Focus on Fast-growing and High-yield Timber Plantations, Key Shelterbelt Development Programs in Such Regions as the Three North and the Middle and Lower Reaches of the Yangtze River, and Biomass Forest Bases Development Program; to strive for the increase of forest area, and increase the capabilities of forests in China to sequestrate carbon; to attach great importance and strengthen sustainable forest management, improve the productivity of per unit forest land, and increase the annual increment of per unit forest area and its capacity to store carbon; to adopt effective measures to strengthen the management and control of forest fires, forest pests and diseases, and wildlife epidemic sources and diseases, rationally control forest resources consumption, and combat destructive harvesting and illegal occupation of forest lands and wetlands, so as to protect forest, desert and wetland ecosystems as well as biodiversity effectively, and reduce forest emission; to actively promote adaptation management measures in the course of forestry production, enhance the capacities of forestry to adapt to climate change, and bring into full play the roles of forestry in the National Strategies to Address Climate Change.

3.2 Periodical objectives ①. The overall objective will be realized in three stages.

3.2.1 By 2010, the area of annual afforestation (including closing mountains for natural regeneration) will be over 4 million hectares②, with the forest coverage of 20% and the forest stock volume of 13.2 billion m^3. Areas with extremely adverse eco-environment, especially

① With reference to relative parts of *A General Review on Sustainable Forestry Strategy Research in China* and *the* 11th *5-year Plan and Middle-and-Long-term Forestry Development Program*

② With reference to the results of the Sixth National Forest Inventory(1999 ~ 2003)

with severe water and soil erosion and desertification at the middle and upper reaches of the Yellow River and the Yangtze River will be under initial control. The protection area of national key public welfare forests will reach 51 million hectares, and 50% of natural wetlands will be effectively protected. The proportion of good seedlings used in plantations will be over 50%. At that time, the capability of forestry to sequestrate carbon will be greatly increased.

3.2.2 From 2011 to 2020, the area of annual afforestation (including closing mountains for natural regeneration) will be over 5 million hectares, with the forest coverage of 23% and the forest stock volume of 14 billion m^3. The area of newly-controlled sandified lands will be over 50% of those suitable for control. About 110 million hectares of national key public welfare forests will be effectively protected. More than 60% of natural wetlands will be taken under sound protection. The proportion of good seedlings used in plantations will be over 65%. It will be accomplished that the forest area in 2020 will be increased by 40 million hectares compared to that in 2005, and the forest stock volume will be increased by 1.3 billion m^3. At that time, the overall functions of forest ecosystem to store carbon will be further strengthened, and the capability of forests to sequestrate carbon will be further improved.

3.2.3 By 2050, the forest area will have a net increase of 47 million hectares compared to that in 2020, and the forest coverage will reach and continuously keep over 26%. Typical ecosystems will be put under sound protection. Sandified lands suitable for control will be controlled in general. Natural wetlands all over the country will be effectively protected, rehabilitated and rationally utilized. Nearly all the plantations in the country use good seedlings. The focus of the forestry development will be transformed to sustainable forest development at full scale. The capability of forests to sequestrate carbon will be kept steadily.

Part 5

Key Areas and Major Actions in Addressing Climate Change in Forestry

In order to bring into full play the unique function of forestry in addressing climate change, and based on the sustainable strategy of national forestry development, the middle and long-term development plan, and the requirements of the *National Climate Change Program* in forestry development, the key areas and major actions in addressing climate change in forestry are determined from the two aspects: climate change mitigation and adaptation.

1 Key Areas and Major Actions in Mitigation of Climate Change in Forestry

Area one: Afforestation

Action 1: Push forward the nationwide voluntary tree-planting campaign. Governments at different level should continue to promote the nationwide voluntary tree-planting campaign according to *the Resolution on Carrying out Nationwide Voluntary Tree-planting Campaign* approved by the National People's Congress (NPC), and *the Measures for Implementing*

Nationwide Voluntary Tree-planting Campaign issued by the State Council. The implementation of the Nationwide Voluntary Tree-planting Campaign should be included in the daily agenda and the leading member at all levels must shoulder a responsibility system. The responsible management system should be clarified and the function of township governments and urban neighborhood offices should be strengthened in organizing the tree-planting campaign. Inspection and supervision of different departments and institutions' implementation of voluntary tree-planting campaign should be conducted to explore and enrich the forms of voluntary tree-planting campaign so as to raise the responsibilities of the entire people. Various departments, institutions and all walks of life should be further motivated to participate in the greening activities, with focus on the greening in cities, green passages, villages and campuses.

Action 2: Implement key afforestation projects to increase forest area. The existing achievements of the Natural Forest Protection Program should be consolidated; the commercial logging in the program area should be restricted as always; the barren land and hills in the program area should be afforested; and the existing natural forests should be comprehensively and effectively protected.

The Conversion of Cropland to Forests Program should be further inspected and checked with policy implementation, forest ownership license, benefit monitoring and post-program management. The supporting policies concerning cropland construction and relative achievements consolidation should be carried out and program evaluation should be conducted in order to push forward the program steadily.

For the Sandification Control Program for the Areas in the Vicinity of

Beijing and Tianjin, afforestation in the barren land and hills and sandification control should be strengthened and advanced technologies and control modes should be promoted. Bans on over-cultivation, over-grazing and over-logging should be strictly carried out. Tending and management of forest stands should be strengthened to effectively consolidate the profits of the control program.

The Three North Shelterbelt Development Program should focus on sand control and water and soil conservation to establish a farmland shelterbelt forest system in the Three North regions. The emphasis should be put on the establishment of regional shelterbelt forests and demonstration sites. By further mobilizing the whole society's participation, we should strive to develop an ecological and economic shelterbelt system as a stable ecological screen in the north region.

Measures for greening projects of shelterbelt forests in the areas including the Yangtze River, the Pearl River, coastal areas, and the Tai Hang Mountain, and plain areas should vary according to different regions' control requirements. For the Yangtze River shelterbelt forest, special efforts should be put on the water and soil erosion control and restoration of low-benefit forests in the Poyang Lake, the Dongting Lake valleys, the Three Gorges and the Danjiangkou Reservoir region to solidify the development achievements. For the Pearl River shelterbelt forest, rocky desertification control should be emphasized with mountains closure for afforestation and forests for conservation of water supply and forests for water and soil conservation as well. Based on the existing forest resources, the coastal shelterbelt forest should be enlarged and improved. The coastal forests should be put more stress on protection, restoration and management of mangroves, while great efforts should be made to restore and protect coastal mangroves and wetlands in all aspects in order to raise the capability to withstand marine disasters and reduce the social impacts and economic losses brought by rising sea level to a minimum. Greening projects in the Taihang Mountain should focus on building ecological screen in the North China plains and establishing and protecting forests for conservation of water supply in riverhead areas. Greening projects in plain areas should lay stress on establishing high-standard cropland shelterbelt forests in North China plains and Northeast China plains, speeding up tree-planting along roads sides and river sides and around villages and courtyards, renewing cropland forest networks in plain areas and carrying out the green passage project efficiently.

For the Forest Industrial Base Development Program in Key Regions with a Focus on Fast-growing and High-yield Timber Plantations, forestry industry companies are encouraged to develop forest material bases with large diameter timber and bamboo forest bases so as to push forward the integration process of forest paper and forest board. Timber industry of fast-growing and high-yield forest bases and national timber reserves are going to be constructed. Provincial project plan is going to be mapped out and implemented to perfect the technical standard of fast-growing and high-yield forests, improve the project quality and gradually strengthen the carbon storage capacity of timber plantations.

Action 3: Quicken the cultivation of rare tree species for timber forest. In the land suitable for afforestation, combining with projects including industrial material forest base, natural forest protection and conversion of cropland to forests, the establishment of timber forest base of rare tree species should be actively carried out. Rare tree species in the natural forests should receive efficient cultivation and sustainable utilization;

forest resources including rare and natural timber forest and forests for other uses should be intentionally cultivated. The cultivation technology of rare tree species for timber forests should be optimized to choose excellent forest types, reasonably distribute the density of forest stand, optimize the structure of forest stand, and increase efficiency for solar energy utilization and the production capability of forest stand.

Area two: Forestry biomass energy

Action 4: Conduct the program of cultivation of energy forests with the integration of processing and utilization. *The National Plan for Energy Forest Development* should be implemented as soon as possible. According to the plan, firstly, marginal lands including mountainous areas and sandification areas and land suitable for afforestation are to be made good use of, with focus on developing woody tree species for producing oil including *Jatropha curcas*, *Pistacia chinensis*, *Xanthoceras sorbifolia*, and *Cornus wilsoniana* and establishing cnergy forest demonstration site for the production of biological diesel oil. We will concentrate our efforts on the joint biomass energy program with China National Petroleum Corporation and Natural Gas Groups. Secondly, we will fully exploit the shrub resources developed from the projects including conversion of cropland to forests and sandification control, and the residues from final cutting, selective cutting and timber processing to manufacture an efficient solid molding fuel for generating electricity or providing heat. Thirdly, we will actively support the development of efficient conversion technology of biomass energy to electricity, directional pyrolysis gasification technology and liquefied oil extraction technology to gradually form " integration of forest and energy" structure to be engaged in material cultivation, processing and production, marketing and sales, science and research.

Area three: Sustainable forest management

Action 5: Carry out forest management program. With increase of annual growth of existing forest as its goal, "Management Plan for Commercial Timber Plantations" will be formulated and implemented. With improvement of forest's ecological function as its goal, "Management Plan for Key National Public-benefit Forests" will be formulated and implemented. On the national and provincial level, the principle of classification management, which requires suitable management policies based on different natural, geological and economic characteristics, and different region and forest types and reasonable designation of public-benefit forest and commercial forest, will be enforced. On the county level, efforts will be concentrated on conducting forest management plan, clarifying different forest types' cultivation goal and management model. On the management organization level, efforts will be concentrated on formulating and implementing forest management plan, carrying out diverse management methods in different sites and completing annual targets. On the forest stand management level, we will give full play to the modern forest management techniques and methods, increase the forestland's productivity to a maximum, and maximize the benefit of each forest stand. While in the process of implementing sustainable forest management project, a set of demonstration sites will be established to explore distinct forest management models in different circumstances, and to popularize guidance on sustainable forest management so as to establish a Sustainable Forest Management (SFM) standard which is in accordance with China forestry development. We will seriously implement technical regulations including *Formulation and Implementation Measures of Forest Management Plan* and *Technical Regulations of Ecological*

Public-benefit Forest Tending in order to strengthen the concept of forest health, continuously improve the stress resistance and stability of forest ecological system and fully tap the potentials of forest resources' carbon sink capability.

Action 6: Expand the area of closing off mountain for natural regeneration and scientifically reform single species plantations. Closing off mountains for natural regeneration is a cost-effective and low-carbon-emission forest restoration method. Thus, we need to expand the area of this activity as large as possible and accelerate the restoration process of secondary forests. We will strengthen the management of existing plantations, and moderately plant broad-leaved species in conifer plantations to gradually solve the problems of "over-dense, over-sparse and over-pure"。We will try our best to avoid planting several generations of coniferous forests on the same site in succession. Based on the climate change in the future, we will try to avoid planting large area of single tree species in the overlapping area of climatic zones, and endeavor to improve the capability of single tree species in withstanding extreme and adverse weather.

Area four: Protection of forest resources

Action 7: Reinforce the logging management of forest resources. We will strictly enforce the forest logging quota, and execute classification management for the public-benefit forests and commercial forests. Forest ecological benefit compensation system needs to be perfected for the public-benefit forest to ensure that forest brings ecological benefit steadily and efficiently. For the commercial forest, especially the fast-growing and high-yield plantations and industrial material forests, we will, in accordance with law, relax cutting control and give priority of meeting the logging quota. We will revise *Management Measures of Forest Logging and Regeneration* and relevant logging regulations to make the utilization and management of forest resources more scientific and legalized. On the basis of scientific zoning, we will apply different logging management model to different regions according to the forestry development and main requirements of forest functions. Through combination of forest logging management with classification management and forest management plan, we can effectively protect and scientifically manage forest resources.

Action 8: Strengthen the requisition and expropriation management of forestland. We will scientifically formulate regional plan of forestry development and national guidelines of forestland protection and utilization plan in order to clarify different regions' forestry development strategy, leading function and productivity layout. We will strengthen the management of forestland protection to place the forestland and cropland on an equal basis. We will adopt the most strict protection measures to establish and perfect quota of management forestland expropriation and impropriation, expert assessment, and pre-evaluation system. We will carry out protection plan of forestland and forestland utilization regulation, and strictly reinforce vegetation restoration system on the occupied and expropriated forestland, to maintain a balance between demands and needs and reduce the carbon emission caused by these activities to a minimum.

Action 9: Improve forestry law-enforcement capability. We will gradually establish a forestry administrative and law-enforcement system with clarified rights and responsibilities, regulated conduct, effective supervision and vigorous protection. The system will make full use of competent forestry departments at all levels and their supporting organizations including forest police, forestry administration inspection teams, timber inspection station, forestry working posts and the vast majority of foresters to protect the forest resources. We will reinforce law enforcement and severely crack down on any illegal acts which destroy the forest resources. We will launch particular campaigns in the areas with serious problems of forest resource destruction and management disorder. Some major and typical cases should be investigated thoroughly and relentlessly.

Action 10: Improve forest fire prevention capability. We will pursue the principle of "prevention first, with active extinguishment" and adopt comprehensive measures to improve forest fire prevention and control in all aspects and reduce forest fire occurrences and costs to a minimum. We will diligently implement *Forest Fire Prevention Regulations*, and strengthen the institutional building of emergency management with forest fire prevention headquarter as its core. We will launch *National Long-and-Middle-term Development Plan of Forest Fire Prevention* to improve fire prevention equipment and infrastructure, to strengthen the building of professional forest firemen, to accelerate the construction of biological firebreak, and to improve emergency response capability of forest fire. We need to strengthen the forecast of forest fire and establish a fire warning system and classification response mechanism. We will adopt a set of prevention measures including increase of publicity of forest fire prevention, fire source management, and hidden fire elimination, to raise people's awareness of fire prevention, and reduce the man-caused fire disaster, thus we can promote the fire prevention work from passive defense, emergent extinguishment to active defense and prepared extinguishment, and realize the goal of "putting out fire completely at the earliest stage, when it is still small". We will enhance research and development capability and introduce new technologies and equipments, gradually expand application scope of large-scaled fire extinguisher facilities including big air tankers, all-terrain carries, to enhance big forest fire extinguishment capability. We will cooperate and negotiate closely with neighboring countries so as to establish an emergency mutual aid mechanism in case of natural fire breakout.

Action 11: Improve prevention and control capability of forest pests, rats, and hares damage. We will adhere to the principles of "prevention first followed by law enforcement, scientifically control and health improvement", and do a good job of preventing and controlling forest pests, rats and hares. *The Regulations on Forest Pests and Diseases Control* is going to be revised. We will strengthen and perfect the elimination of invasive alien species including nematode disease, American white moth, *Brontispa longissima*, *Dendroctonus valens*, pine scale, poplar wood borer and harmful native pests and diseases. 2008 ~ 2015 Plan of Harmful Forest Creature Control is going to be formulated and carried out. We will strengthen all aspects of monitoring and forecast of forest pests, rats, and hares and the construction and management of 1000 national monitoring centers. We will reinforce cooperation with national meteorological department to make the forecast more scientific, timely and accurate. We will strengthen cooperation with customs department in the inspection and law enforcement of forest pests and diseases and be strictly on guard against the invasion of alien species.

Area five: Forestry industry

Action 12: Reasonably develop and utilize biomass material. We will make sure that biomass material and bio-pharmaceuticals are developed and utilized reasonably and hammer out development and utilization plan of biomass materials. We will put into effect *the Main Points of Forestry Industry Policy* to avoid low-level repetition, and restrict the number of high pollution and high energy consumption companies and thus to promote the development of sound forestry economic circulation. On the basis of consolidating traditional use of timber, through characteristic improvement of wood products, the application of timber is widely extended in the fields of construction, packing, transportation and energy, so that the wooden construction timber will be vigorously developed. In places where land resources are relatively rich such as rural areas, suburbs and scenic spots, we will promote the construction of wooden houses at full tilt. We will forcibly develop high-quality wood-based panel. Application of timber, especially bamboo wood in fields of doors and windows construction, wall materials, construction model and containers' baseboard is going to be expanded. We will widen the application scope of timber products, and moderately advocate partial substitution of timber products for fossil fuels.

Action 13: Improve the efficiency of timber recycled utilization. We will vigorously promote the principle of "energy saving, low consumption and emission reduction" in timber industry and high efficiency and recycled use of timber resources. We will forcefully develop fine processing of timber and deep processing industry. We will adopt different processing techniques according to tree species, age and different parts of wood to make best use of timber. Logs and the residues of logging, cross cutting and processing are going to be utilized with new technologies to produce wooden reconstituted timber and wood-based composites. We will actively develop timber preservation industry to speed up the industrialization of timber preservation and wood modification of plantations, realize standardization and series of timber preserving products and gradually establish and perfect the inspection and examination system of timber preserving products and thus to improve timber performance and prolong the service life. In accordance with *the Main Points of Forestry Industry Policy*, we will timely formulate implementation rules and regulations of timber industry. For the restricted programs, concrete requirements and restrictions of industry access are going to be raised; for the eliminated programs, forceful measures will be taken to eliminate them in specified time. A set of laws, regulations and standards including regulations on integrated utilization of timber processing resources, national standard on comprehensive utilization of timber and regulations on energy saving and low consumption in timber industry to promote high efficiency and recycled utilization of timber should be worked out. We will vigorously push forward clean production of timber industry companies and recycled use of resources. Certification of clean production, quality, and environment will be strengthened. Supervision of high efficiency and recycled use of timber and products identification management will be reinforced. Green environment logo and market access system will be established to curb production with high energy consumption, high pollution and low benefit from headstreams.

Area six: Wetland restoration, conservation and use

Action 14: Launch rescuing conservation and restoration in important wetlands. We will concentrate major efforts on solving ecological supplement of the missing water in important wetlands. We will carry out pollutants control work in a planned way, launch the project of converting cropland to wetlands to broaden wetland areas and improve wetlands ecosystem quality. Based on different conditions including wetland types, degradation causes and degrees, vegetation recovery on wetlands will be conducted to suit local situations to improve wetlands carbon storage.

Action 15: Develop demonstration sites of sustainable agriculture, animal husbandry and fishery development. We will establish national demonstration sites of comprehensive utilization for agriculture, animal husbandry and fishery, wetlands protected and management areas for agriculture, animal husbandry and fishery, southern man-made demonstration wetlands of high-efficiency ecological agriculture, and demonstration bases of wise use of mangrove wetlands to optimize breeding industry in coastal areas. The implementation of ecological breeding will improve domestic sustainable use of wetlands in agriculture, animal husbandry and fishery, and reduce greenhouse gas emission cause by wetlands destruction.

2 Key Areas and Major Actions in Adaptation of Climate Change in Forestry

Area one: Forest ecological system

Action 1: Improve adaptability of plantation eco-system. While respecting natural and economic principle, based on future climate change estimated situations, we will make scientific plans and decisions on national afforestation area and choose suitable forest category and afforestation tree species, especially those good indigenous and fire-resistant species; we will vigorously plant mixed tree species forests and mixed broadleaved and conifer forests to establish a plantation eco-system with high level of stress resistance and adaptability. At the same time, we will combine afforestation technologies with forest fire prevention to reduce forest fire occurrences. Full consideration should be taken to long-and-short-term carbon fixation effects of standing forest, and we will choose intolerant tree species with its growth under full sunlight and shade-endurance tree species scientifically to form multi-storied forests and uneven-aged forest stands. In arid and semi-arid areas, we will prevent and control desertification with steady steps, in line with local conditions, make great efforts to plant trees and grasses and conduct mountain closure for natural regeneration and grass growing. We will reasonably regulate the flow of water resources for ecological use, establish and consolidate the ecological shelterbelt with forests and grasses as its main body. We will strengthen management of plantations, increase the general function of ecosystem of plantations and conserve biodiversity. We will further expand the management scope of water and soil erosion with biological measures, to reduce organic carbon losses due to water and wind erosion.

Action 2: Establish nature reserves of typical forest species. On the basis of existing nature reserves, the nature reserves of typical forest species will be established for the forest ecological systems with small and narrow-distributed area in different climate zone and without conservation of nature reserves or the conservation ratio is relatively low. We will focus on bringing extremely endangered and single and terrestrial wildlife species and their habitats into the nature reserve, and priority will be given to the terrestrial wildlife with relatively less amount, more limited restricted distribution and severely segmented habitats. We will integrate nature reserves of one biological and geological unit with identical or similar ecosystem types in the light of principles of uniformed planning, management and administrative district zoning in order to establish a network of conservation, ensure complete function of the eco-system and improve conservation efficiency of the nature reserves.

Action 3: Intensify the conservation of key species. Firstly, conservation of key species should be conducted according to their distinctiveness and conservation coverage, and the different on the spot conservation measures be adopted in an order of priority. Secondly, the establishment of nature reserve should be precedent for the species which indeed need to be rescued on site, and their habitats or their original habitats which have not yet been included in the network of nature reserves; For the places where it is incapable of establishing standard nature reserves, small protected areas should be set up and managed by neighboring national nature reserves. For the species with large population excluded from conservation network, the scope of nature reserves should be moderately expanded to include entire or most of the habitats of the population, or for the local nature reserves which focus on protecting key species, when conditions are ready, they can be promoted as national nature reserves. The habitats of relatively small population of species in national nature reserves should be improved and broadened. Thirdly, several national nature reserves should be selected as core zone to protect key species on site and the species with concentrated distribution. In the areas devoid of conservation, new nature reserves should be established according to the local conditions to expand the conservation ratio of species and their habitats. For the wildlife which are migratory species or with large home range, nature reserves should be established on their habitats and activity passages, and special attention should be given to the connection of different nature reserves. Fourthly, as to the species with strong dependence on habitats, we will strengthen their ecosystem conservation in order to protect, restore and expand species and their habitats. Fifthly, for the species with broad distribution area, we will select nature reserves with good conditions accordingly as developing priority reserves to improve nature reserves' level. Sixthly, for those species which have already distributed in the national nature reserves, we would increase capital input to improve, restore and expand their habitats.

Action 4: Improve wildlife epidemic-stricken area and disease monitoring and pre-warning capability. We will adhere to the principles of "intensified leadership, close cooperation, science orientation, control by law, prevention and control by the people, and resolute disposition" to do a good job of epidemic-stricken area and disease monitoring and prevention of wildlife. We will further strengthen and improve monitoring system, emergency management, and conduct background investigation of wildlife epidemic diseases. Monitoring and pre-warning of wildlife epidemic diseases and construction and management of national monitoring station will be strengthened in all aspects. We will tighten the cooperation with hygienic and agricultural departments to form an unified prevention and control mechanism. We will give attention to the personnel training and get ready for emergency practice.

Area two: Desertification ecosystem

Action 5: Strengthen vegetation protection in desertified areas. A batch of nature reserves will be set up in the desertification ecosystems of the riverheads where semi-arbor trees, shrubs, semi-shrubs and small cushion semi-shrubs are severely damaged by human behaviors. While we are protecting the distinctive vegetation resources in the western part of China, and heighten their adaptability to climate change, we will also strengthen carbon sequestration of the ecosystem to improve the fragile ecological environment in the western areas. The development priority will be stressed on the desert vegetation areas where national nature reserves have not yet been established, especially the desert vegetation types with small and restricted distribution areas, for example, the *Calligonum leucocladum* desert, should be totally brought into nature reserves.

Area three: Wetland ecosystem

Action 6: Strengthen fundamental work of wetland conservation. We will conduct the second national wetland resource survey to fully understand the features of China's wetland resources, social and economic situation, main challenges and development trends. We will investigate and evaluate carbon sequestration amount and capability of the wetland eco-system. The second national peatland resource survey will be carried out as soon as possible to size up the resource distribution, amount, conservation, development situation and trends.

Action 7: Establish and improve network of wetland nature reserves. We will strengthen development of existing wetland nature reserves, and improve infrastructure, establish sound management institutions and launch community co-management in line with *the National Plan for Wetland Conservation Project* and *the National Implementation Plan for Wetland Conservation* (2005 ~ 2010). We will focus on conserving different wetland types including seaside wetlands, marshes and peatland, and develop wetland parks to suppress the trends of wetlands shrinkage and degradation so as to form a wetland natural protection network with a relatively comprehensive conservation and management system.

Part 6

Safeguard Measures

1 Improve Leadership to Take Active Forestry Actions against Climate Change

Forest departments at all levels should increase the awareness of the special status and role of forestry in addressing climate change and combine the work of forestry with the addressing of climate change. It is required to practically strengthen the organization and leadership, seriously fulfill obligations, actively work out measures and combine the reality of localities to virtually implement various forestry measures and achieve the targets of addressing climate change and authentically give play to forestry in the mitigation of and adaptation to climate change.

2 Intensify Science and Technology to Promote Forest-related Scientific Researches on Climate Change

Research should be done aiming at the forestry tasks and main actions determined by the national strategy in addressing climate change. Firstly, we should carry out in-depth basic research on forests' response to climate change to keep abreast of international research and hold thorough discussions on the circulation processes and coupling mechanisms of carbon, nitrogen and water in forests, wetlands and desert ecosystems in the scenario of climate change. We should also study on the mechanism of human activities' impacts on the carbon source/sequestration functions of forests, wetlands, deserts and so on. Secondly, combining· the characteristics of the geographical distributing areas of forests and ecological environment types in China, we should reinforce the planning and development of forest ecosystem positioning stations and improve the located field observations on forest ecosystems' response to climate change. Through research on located field observation technologies for biodiversity, forest fires, pests and diseases, the forest ecosystem observation network and monitoring system will be gradually improved, based on which studies on policy proposals, technical options, cost benefits and assessment of forests' adaptability to climate change will be strengthened and forestry's adaptability to climate change will be improved. Thirdly, the research on forest carbon inventory and monitoring system should be enhanced and a corresponding national system should be formed as early as possible for data sharing. Fourthly, technological research on reducing emission and increasing forest carbon sequestration should be intensified. Cooperative research on efficient cultivation of bio-energy forest plantations, extraction of biological diesel oil, ethanol produced from cellulose biomass, power generation by biomass, etc. will be carried out. The analysis on the emission-reduction and carbon-sequestration-increase potentials and cost benefits of key programs and areas will be carried on, with research focused on arrangement pattern of forest types and tree species, cultivation and harvest regimes with attention to multiple benefits, rehabilitation and restoration technologies for wetlands, mangroves and other ecosystems, sustainable management technologies, management technology for agro-forestry systems, etc., from the perspectives of carbon sequestration, wood supplies, water source conservation, and habitat protection. Fifthly, study on a forest disaster-prevention system should be intensified. The occurrence mechanism of forest fires, pests and diseases in the scenario of climate change should be studied and the research and development and introduction of new fire-prevention and control techniques and equipment should be reinforced. Impacts of major forest pest and disease disasters will be identified and countermeasures for prevention and control will be put forward. Studies on pathogenic mechanism of various wildlife epidemic diseases in the context of climate change especially the zoonosis should be intensified to gradually understand the pathogenesis of wildlife epidemic diseases and master key techniques on rapid diagnoses and examination. Sixthly, studies on issues related to the adaptability of forests, wetlands, deserts, urban green spaces and other ecosystems in the scenario of climate change should be strengthened so as to put forward technical solutions to adaptation, and related analysis and evaluation of costs and benefits should be carried out.

3 Enhance Training to Improve Capacity Building for Forest Practitioners

Firstly, we should strengthen the expert team building for the response to climate change and actively encourage middle-aged and young scientists to do scientific research in forest-related fields to respond to climate change through scientific research programs and fair-play mechanisms and other measures to gradually foster a number of experts on climate change with political quality, high research capability and down-to-earth working ethics. Secondly, the training for related personnel on forestry's response to climate change should be reinforced. Such training should be closely combined with the implementation of *the National Working Program of*

Forestry Talents for the Eleventh Five-year Period and the Medium and Long Term, *the Outline of National Action Plan on Forestry Practitioners' Scientific Quality*, the Program on Knowledge Updating of Forestry Professionals and so on. The forestry practitioners should retain conceptions of sustainable development, resources saving, ecological protection, environment improvement, reasonable consumption, circular economy, etc. Thirdly, the preparation of *the Outline of Local Leadership Training on Forestry's Response to Climate Change* should be well organized. The relationship between ecosystems and climate change, the responsibilities and role of local governments in strengthening ecological improvement and protection and responding to climate change, forest-related measures to respond to climate change and so on should be included as the important contents of the thematic training on forestry for local leadership, and related training courses and teaching materials should be developed. Fourthly, we should hold more thematic workshops and lectures on forestry's response to climate change and include related contents in various official and employee training programs.

4 Deepen Publicity to Continuously Raise the Public Awareness of Climate Change

Firstly, we ought to extend the science publicity and popularization to deepen the whole society's understanding of the role and functions of forestry. Knowledge on forest cultivation and management and the role and functions of forests and wetlands to increase carbon sequestration by absorbing and store carbon should be publicized and extended to raise the public awareness of forestry as an economically effective and important measure to mitigate the warming and its irreplaceable role in addressing climate change, to urge the whole society to review the status and role of forestry, increase the public awareness of ecological and climate protection, and encourage more people to join in the action of "planting trees to increase carbon sequestration, protecting forests to stabilize carbon, improving environment and addressing climate change". Secondly, the dynamic publicity should be well implemented. The important forestry measures and specific actions in addressing climate change as well as the multiple benefits brought by such measures in increasing forest carbon sequestration, preventing soil and water erosion, controlling desertification, alleviating poverty and protecting biodiversity should be well publicized. We should encourage people to pay further attention to forests via various media and approaches and publicize the response of forestry to climate change in an all-round way and from various perspectives. Thirdly, the publicity of models should be enhanced. Models of ecological improvement in various localities should be highlighted and the changes and experience of key areas in advancing tree-planting and improving ecological conditions should be reported. Actions of social groups, organizations and key enterprises at all levels in tree-planting, afforestation, emission reduction and carbon sequestration increase should be given publicity in a big manner based on the China Green Carbon Fund. A series of model examples of enterprises, groups, organizations and individuals joining forestry actions in addressing climate change should be selected and publicized to drive the whole society to take actions.

5 Innovate Mechanism to Accelerate the Forestry Reform and Response to Climate Change

Firstly, we should thoroughly implement *the Opinions of the Central Committee of the Communist Party of China and the State Council on Boosting the Reform of Collective Forest Tenure System in an All-round Way* by clarifying property rights and improving policies to provide farmers more rights of development, strengthen their ability to develop, guarantee their legitimate rights and interests and further mobilize the vast forest farmers to establish, protect and manage forests. Secondly, enormous support should be provided to social entities of various kinds for their participation in forestry development. Guided by policies and guaranteed by legislation, societies, enterprises, foreign businesses, etc. should be encouraged and supported to establish forests in various ways to continuously enlarge the number and scale of non-public-owned plantations. Thirdly, laws and regulations on forest protection will be formulated or revised and related matching policies will be improved to promote forestry development with legislative and policy support. The *Forest Law* should be revised in time and the national statutes on wetland protection should be issued as early as possible. The law enforcement will be intensified and the social supervision will be extended. Fourthly, the responsibility system for tree-planting target management by governments at various levels and the responsibility system of sectoral tree-planting should be improved continuously and more forms of nationwide voluntary tree planting in the context of market economy should be further studied to advance the in-depth development of voluntary and sectoral tree planting. Fifthly, China Green Carbon Fund can be fully used as a platform and enterprises, organizations, groups and individuals should be encouraged to join in the Fund. Qualification management should be done on carbon inventory, monitoring, registration, etc and the certification system for qualifications will be gradually set up.

6 Underline Priorities to Increase Funding for Forestry Development

Firstly, the continuous capital investment in forest development and protection should be maintained under the public financial system. According to the fields of priorities and principal actions of forests in addressing climate change, the funding support by public finance should be leveled and enhanced for the afforestation of key programs, sustainable forest management, valuable species cultivation, the prevention of forest fires, pests and diseases, the monitoring, prevention and control of wildlife epidemic diseases and causes, forest protection and so on, in order to ensure the effective implementation of such actions. Secondly, funding for forest development and protection should be raised from various sources; the program plan to use foreign investments in forestry should be systematically organized and well formulated to introduce foreign investments into key forestry programs; and national policies and management system for forestry credit capital investments should be further improved. Thirdly, the efficient recycled utilization of wood will be supported and the national support will be sought for enterprises to ameliorate their technology in improving the efficiency for integrated use of wood. Fourthly, we will increase the funding support for special scientific research on technologies of forest carbon inventory and monitoring, forest restoration, afforestation on harsh site conditions, integrated sustainable management, forest adaptability evaluation, capacity building, publicity and training, implementation of international conventions, which are parts of the solutions of forestry to climate change.

7 Extend International Cooperation in Forestry while Taking the Interests of All Sides into Account

Firstly, taking the Asia-Pacific Network on Forest Rehabilitation and Sustainable Management (APFNet) as a platform, the regional and international cooperation in forestry will be actively developed. Secondly, we will participate in all negotiations on forest-related topics in the international processes under UNFCCC and the *Kyoto Protocol* in an in-depth way, organize experts inside and/or outside the forestry sector to study on the negotiation countermeasures for forest-related topics, and support Chinese forestry experts to join IPCC and related work. Thirdly, we will actively boost the afforestation for carbon sink under CDM to accumulate experience of project implementation and study on approaches to promote the afforestation drive within China by taking advantage of international mechanisms. Fourthly, we encourage the bilateral and multilateral cooperation and dialogue mechanisms on forestry and climate change. We will use resources of international cooperation to develop and improve the capacity building for forestry in addressing climate change and encourage the transfer of advanced management concepts and techniques from developed countries. We will seek to increase our cooperation with developing countries in climate change through foreign aid channels.